主编 卢一 杜莉

中国川菜

（中英文标准对照版）

SICHUAN (CHINA) CUISINE IN BOTH CHINESE AND ENGLISH

四川科学技术出版社

图书在版编目（CIP）数据

中国川菜：中英文标准对照版/卢一，杜莉主编. —成都：四川科学技术出版社 2014.2（2023.11重印）
ISBN 978-7-5364-6964-8

Ⅰ.①中… Ⅱ.①卢… ②杜… Ⅲ.①菜谱－四川省－汉、英 Ⅳ.①TS972.182.71

中国版本图书馆CIP数据核字（2009）第241889号

中国川菜
（中英文标准对照版）

Sichuan (China) Cuisine in Both Chinese and English

主　编　卢一　杜莉

出 品 人	程佳月
责任编辑	程蓉伟
装帧设计	程蓉伟
责任出版	欧晓春
出版发行	四川科学技术出版社
地　　址	四川省成都市锦江区三色路238号新华之星A座 传真：028-86361756　邮政编码：610023
电脑制作	成都华桐美术设计有限公司
印　　刷	成都市金雅迪彩色印刷有限公司
成品尺寸	210mm×285mm
印　　张	20
版　　次	2014年4月第2版
印　　次	2023年11月第4次印刷
书　　号	ISBN 978-7-5364-6964-8
定　　价	198.00元

■ 版权所有·翻印必究 ■

地址：四川省成都市锦江区三色路238号新华之星A座25层
邮购电话：86361770　邮政编码：610023
如需购本书请与本社发行部联系

中国川菜（中英文标准对照版）
编辑委员会

编委会顾问
廖伯康

编委会主任
钟 勉　陈光志

编委会副主任
袁本朴　于 伟　吴果行

编委会委员
陈建辉　薛 康　刘云夏　张 谷　崔志伟　秦 琳
卢 一　李 新　李树人

主编
卢 一　杜 莉

副主编
陈祖明　麦建玲

英文主审
Fuchsia Dunlop［英］　Shirley Cheng［美］
卢 一　王维民　林 红

中文撰稿
陈祖明　尹 敏　罗 文　彭 涛　杜 莉　张 茜

英文翻译
张 媛　潘演强　郑贤贵

菜点制作
陈祖明　陈应富　江祖彬　尹 敏　罗 文　熊 军　辛松林

菜点摄影
李 凯　李云云

参编单位
四川省旅游局
四川省外事办公室
四川省对外友好协会
四川烹饪高等专科学校
四川省美食家协会

"食在中国，味在四川。"川菜作为中国久负盛名的地方风味菜系之一，历史悠久，声名远播，在我国烹饪史上占据了重要地位，业已成为中国餐饮文明的重要组成部分。川菜发源于我国西南地区的四川盆地，距今已有两千多年的历史。中国川菜不仅养育了勤劳智慧的四川人，而且还传承着四川民俗文化和中国文化。如今，中国川菜正以其独特的魅力，充分展现出取材广泛、调味多变、技法多样、品类繁多和清香醇浓并重、麻辣鲜香见长的鲜明特色。川菜经过跌宕起伏的演变，历经岁月风霜的锤炼，在全国已呈燎原之势。祖国大地，处处皆闻川菜之香！五湖四海，洲洲都见川菜之美！川菜正在被越来越多的人所接受、所喜爱。

随着中国经济的发展，川菜已经成为四川特色鲜明的旅游资源和重要的支柱产业之一，更多的外国友人、境外游客开始认识和喜爱川菜。四川省委、省政府和社会各界也为川菜的国际化发展提供了不少有利的政策支持和制度保障，积极为川菜的繁荣搭建平台、增添助力，从而使川菜走出国门的步伐更加稳健、快捷。但不能忽视的是，川菜产业在快速发展中同样遭遇了一些"尴尬"，明显表现在川菜菜名的翻译长期以来比较混乱、没有统一的标准，在一定程度上增加了国外人士品鉴川菜的难度，无形中成为了制约川菜在海外发展的一道"瓶颈"。

四川省委、省政府高度重视这一问题。在一次交谈中，蒋巨峰省长得知我十分关注和支持川菜产业的发展、还曾主编过《川菜文化研究》文集，便委托我组织力量解决川菜规范化翻译问题。受省长重托，虽耄耋之年，自不敢懈怠，遂组织四川烹饪高等专科学校和四川省美食家协会的精干队伍致力于这项工作。四川烹饪高等专科学校，是邓小平同志亲自关怀下于1985年成立的中国唯一的公办烹饪高等学校，作为有8000多名在校学生的全球最大的烹饪学校，长期开展川菜的教学、研究、创

新和国际交流，成效显著。四川省美食家协会则是中国唯一的省级美食家协会，拥有众多的川菜烹饪大师，美食家和名师、名厨、名店，为推动川菜的发展和广泛传播做出了突出贡献。两者结合，优势互补，组成了一支高效率的工作组，为完成这项光荣而艰巨的任务奠定了良好的基础。

2008年下半年，"经典四川菜点菜名翻译"项目正式启动。首先通过报纸、网络由大众投票推荐，再结合烹饪专家的评审，最终确定了180道经典四川菜点。其中，既有传统精华，又有创新美味；既有高端名品，又有大众佳肴，琳琅满目，多姿多彩。工作组又组织川菜烹饪和英语翻译专家完成了菜点中、英文稿的编写，并邀请国外著名烹饪专家，如英国的扶霞女士、美国烹饪学院的成蜀良教授等审查和修改。与此同时，工作组还组织多位川菜烹饪大师、名师和专业菜点摄影师精心制作、拍摄了180道菜点。可以说，最终摆在读者面前的这部《中国川菜》，凝聚着许多人的辛劳和智慧。

《中国川菜》重点介绍了180道经典四川菜点的制作方法，全书采用中英文对照的编排方式，是目前国内第一本大型的中英文标准版的地方风味精美图文集。本书的制作团队囊括了国内外饮食文化研究领域的一流专家、川菜烹饪大师、摄影名师等，由此确保了本书内容的权威性。本书文字简洁明了，图片美观精致，翻译标准规范，特别是180道经典四川菜点的制作方法，可供感兴趣的中外读者亲自操作实践，因而本书具有很高的可读性、观赏性、实用性和指导性。

我今年已85岁了，但对四川这片我长期生活的土地依然充满着热爱，也深爱着养育四川人民的川菜，能够参加这样一个有意义的项目，我十分欣慰。美哉，川菜！

是为序。

二零零九年九月

（注：作者为政协四川省第六届委员会主席）

Preface

"While China is an ideal place to feed your appetite, Sichuan is the best place to satisfy your taste buds." Sichuan cuisine, one of the most renowned local flavors, has a long history and a high reputation. It originated in the Sichuan Basin in the southwest of China over two thousand years ago, and is now characteristic of a wide range of ingredients, varied ways of seasoning, diverse cooking techniques and multiple categories. While having both delicate and savory dishes that mix different tastes, Sichuan cuisine is especially good at preparing hot, tingly and aromatic food. Despite ups and downs during its course of development, Sichuan cuisine has made its influence felt around China and even around the world and is accepted and loved by an increasing number of people. As a significant school of Chinese cuisine and an inseparable part of Chinese culinary culture, Sichuan dishes are not only a reflection of Sichuan folk customs but a carrier of Chinese culture.

With the economic development of China, Sichuan cuisine has become one of the tourism resources and principal industries of Sichuan. More and more foreigners have come to know and love Sichuan food. The favorable policies of CPC Sichuan provincial committee and Sichuan provincial government as well as the support of different walks of life have helped to speed up the development of Sichuan cuisine around the world. However, we cannot ignore the bottleneck of its development: the lack of standards in the translation of Sichuan dishes hinders foreigners from getting acquainted with Sichuan food.

CPC Sichuan provincial committee and Sichuan provincial government attach great importance to this problem. Governor Jiang Jufeng entrusted me with the task of standardizing the translation of Sichuan dishes, knowing in a conversation with me that I had been concerned about and supportive of the development of Sichuan cuisine and that I had been in charge of the compiling of a book of collected essays entitled Research on Sichuan Culinary Culture. Although in my eighties, I then committed myself to the task without delay. Under my supervision, Sichuan Higher Institute of Cuisine and Sichuan Gourmet Association organized a working team to work on this project. Sichuan Higher Institute of Cuisine, the only state-run culinary college, was founded in 1985 under the guidance of Mr. Deng Xiaoping. Now there are over 8,000 students studying here. The Institute has achieved remarkable results in terms of teaching activi-

ties, academic research, innovation and international exchange. Sichuan Gourmet Association, the only gourmet society at the provincial level in China, consists of masters of Sichuan cuisine, gourmets, renowned chefs, reputed cooks and noted restaurants, and has done a lot to promote the development of Sichuan cuisine. The cooperation of the Institute and the Association set the stage for the successful accomplishment of the task.

The project "Classical Sichuan Dishes and their Translation" was formally launched in the second half of the year 2008. The one hundred and eighty classical dishes chosen through newspaper voting, internet voting and experts' proposals incorporate various categories: traditional or innovative, exclusive or popular. Experts in Sichuan cuisine and translation then set out to compile and translate the recipes. Foreign specialists in this field, such as Ms. Fuchsia Dunlop from Britain and Professor Cheng Shuliang from Culinary Institute of America, were also invited to go through the English translations of the dishes. Meanwhile, masters of Sichuan cuisine, renowned chefs and professional photographers of dishes were responsible for the preparation and shooting of the 180 dishes. The book has been made available thanks to the efforts of these people.

The book *Sichuan Cuisine*, primarily an introduction to the preparation of 180 classical Sichuan dishes in both Chinese and English, is the first major attempt to standardize the preparation processes of local dishes and their translation with beautiful pictures. The working team made up of experts, masters of Sichuan cuisine and famous photographers ensures the authority of this book. The concise wording, exquisite pictures, accurate translation and the detailed preparation processes make it easy for readers at home and abroad to try cooking the dishes by themselves. It is a great pleasure to read the book, appreciate the pictures and prepare the dishes according to the recipes.

I am eighty-five years old and have entertained a deep love for this land and its dishes. How fascinating Sichuan cuisine is!

That is what I want to say for the book.

Liao Bokang

September, 2009

Note: *The writer of the preface was the Chairman of the Sixth Committee of Sichuan Political Consultative Conference*

前 言

四川烹饪高等专科学校和四川省美食家协会十分荣幸地承担了四川省政府委托项目——"经典四川菜点菜名翻译"的组织和完善工作。该项目旨在对经典四川菜点的制作方法和英文翻译进行规范，以便让更多的外国人能够了解甚至制作川菜，从而促进川菜的国际化发展。为了圆满完成该项目，两家单位强强联合组建了一个由精兵强将为骨干的工作组，着力完成了大量而细致的工作。

川菜成千上万种，如何确立经典川菜是摆在我们面前的重要问题。为此，工作组特别邀请了四川省内的著名饮食文化专家、川菜烹饪大师、美食家及其他相关人士经过认真遴选，先期列出推荐名录。这个名录中，既包含有传统川菜，又包括改革开放以来的创新川菜及部分近年来的市场热销川菜；既有五星级酒店的高档川菜，又有适合乡村小店和家庭制作的大众川菜。为了集思广益，工作组还将名录登载于报纸和网络上，由读者和网民投票推选。工作组根据得票高低和专家推荐，最后确定了180道经典四川菜点。

在经典四川菜点的入选名录确定之后，四川烹饪高等专科学校川菜发展研究中心便组织相关专家编写了这些经典四川菜点的制作方法、特点、典故等内容的中文稿，并组织川菜大师、名师以及四川烹饪高等专科学校的教师、专业菜点摄影师积极开展入选菜点的制作及图片拍摄工作，四川烹饪高等专科学校的外语教师同期开始进行中文稿的英文翻译。在翻译过程中，翻译人员遇到了三大困难：一是菜名难以用简短英文准确表示其义，如"宫保鸡丁"中"宫保"是清朝官名，但菜名不可能用一长串词作解释；二是川菜烹调方法在英文中无对应词汇，如川菜的煮、卤、熘，在英文仅有"boil"一词，英文不易准确表示这些差异；三是由于四川地方饮食文化和语言习惯，英文难以表达其含义，如川菜的"𤆵"，不仅英文无对应词，即使四川、重庆以外地区也无此字，就连以前的汉语字典中也没有收录此字。此外，由于以前的翻译者大多没有在英、美等国生活的经历，并不十分熟悉外国人的生活方式及语言表达习惯，因此，以往的川菜英文翻译大多有词不达意的现象，以致于英语国家的人常常看不懂川菜的英文书。为此，工作组做了两件事：一是邀请在英、美两国生活和工作的两位专家仔细审查、修改英文稿，一位是美国烹饪学院（Culinary Institute of America）的成蜀良（Shirley Cheng）教授，她曾在四川烹饪高等专科学校从事川菜烹饪教学近十年，后到美国读硕士并在美国烹饪学院任教二十年；另一位是英国BBC电视台的专栏作家扶霞女士（Fuchsia Dunlop），她曾在四川大学学习汉语，也在四川烹饪高等专科学校学习川菜制作，在成都生活了多年，回国后仍然坚持每年到四川来了解川菜的发展状况，可以说是"川菜通"。二是将翻译稿送给美国驻成都总领事馆的相关人士，请他们提意见。经过反复讨论修改，最终形成了现在的翻译定稿。

作为"经典四川菜点菜名翻译"项目的最终成果，这部《中国川菜》主要包括川菜概述、川菜特色调味品、180道经典四川菜点的制作方法等内容。全书文字均为中英文对照，并配有精美图片，是目前国内第一部大型中英文标准对照版的地方风味菜系精美图文集。它既有很高的文化与艺术价值，又具有很强的实用操作价值。我们深信，这部《中国川菜》一定能够成为一个良好的媒介，为川菜更好、更快地走向世界，并对促进中外饮食文化交流发挥重要的作用。

编　者
2010年初春

Introduction

Sichuan Higher Institute of Cuisine and Sichuan Gourmet Association feel greatly honored to have been entrusted with the project of "Classical Sichuan Dishes and Their Translation". This project aims to standardize the preparation processes of classical Sichuan dishes and their translations so as to provide easier access for foreigners to get to know and even practice Sichuan cuisine, which will then serve to boost the development of Sichuan cuisine around the globe. To achieve this objective, the Institute and the Association jointly set up a working team fully committed to this both extensive and intensive work.

To start with, how were "Classical Sichuan Dishes" chosen with the vast variety of Sichuan cuisine? The culinary culture experts, masters of Sichuan cuisine, gourmets and other specialists in this circle were invited to suggest lists of worthy dishes. Their recommendations incorporated traditional dishes, dishes innovated after China's adoption of the policy "opening-up and reform", dishes popular in recent years, dishes prepared in five-star hotels, and dishes made at home or in small rural restaurants. These long lists were then published in newspapers and on the internet. Readers and netters were asked to vote for the dishes that they thought classical. The working team finally came up with one hundred and eighty dishes after considering the results of the vote and suggestions from the experts.

The Sichuan Cuisine Development Research Center of Sichuan Higher Institute of Cuisine then started with the Chinese version by assigning the dishes to different experts and asking them to compile preparation processes and features of each and to provide background information if necessary. Masters of Sichuan cuisine and teachers at the Institute were responsible for the cooking and shooting of these dishes. Meanwhile, teachers in the Foreign Language Department of the Institute took up the English translation of the dishes. Three elements stood in the way of the translation. Firstly, it is not easy to transfer the cultural connotations of the names of some dishes into their English counterparts in a brief and accurate way. For example, Gongbao in the famous dish Gongbao Diced Chicken refers to a title of an official in the Qing Dynasty, and it is unlikely to fully expound its meaning without using more than one or two words. Secondly, some cooking methods have no counterparts in English. For example, the cooking methods of zhu and lu, though all conveying the meaning of boiling, are totally different in Sichuan cuisine. Thirdly, the culinary culture and language of Sichuan have given rise to some characters that are peculiar to Sichuan. Some of these characters are missing in dialects outside Sichuan, let alone in English. Besides, most translators of Sichuan dishes before either lacked profound knowledge in cooking or had no experience of living abroad or needed further training in translation, which have led to the misleading or incomprehensible translations of some dishes. In view of this, two experts living in Britain and America respectively were invited to go through the translations. One of them, Professor Cheng Shuliang, obtained a master's degree in America and then worked as a teacher in Culinary Institute of America for twenty years after having taught Sichuan cuisine at the Sichuan Higher Institute of Cuisine for nearly ten years. The other expert, Ms. Fuchsia Dunlop from Britain, has been working as a columnist for BBC. Ms. Dunlop, who once lived in Chengdu for years studying Chinese at Sichuan University and cooking at Sichuan Higher Institute of Cuisine, comes to Sichuan every year to get up-to-date information about the development of Sichuan cuisine. In addition, the translations were also presented to the American Consulate in Chengdu to seek their advice. After repeated revision and modification, the English translations of classical Sichuan dishes are now settled.

The book, *Sichuan Cuisine*, is the results of the project "Classical Sichuan Dishes and Their Translation". Incorporated into this book are a brief introduction of Sichuan cuisine, featured seasonings, preparation processes of 180 dishes and exquisite pictures. The book, as the first comprehensive attempt to introduce local food in both Chinese and English, is of immense cultural, artistic and practical significance and is sure to function as a medium to help Sichuan cuisine gain world fame and to play a vigorous role in promoting the culinary culture exchange between China and foreign countries.

川菜概说

中国美食名扬天下，深受世人喜爱。历史悠久的川菜可谓是中国最著名、最具特色的地方风味菜系之一，一直有"食在中国，味在四川"的美誉。

川菜发源于古代的巴国和蜀国，在经历商周至秦的孕育萌芽期后，至汉晋时已具雏形。延自隋唐五代，川菜得到了蓬勃发展。两宋时，四川风味已进入了京都的饮食市场。明代，川菜平稳发展。清朝中期，由于运用辣椒调味，在继承巴蜀时期"尚滋味"、"好辛香"的调味传统基础上，川菜有了进一步发展。晚清以后，川菜逐步形成了充满个性的风味体系。新中国成立后，川菜也同其他地方菜系一样进入繁荣创新时期，尤其是改革开放以来，川菜在保持特色的前提下中外兼收、改革创新，快速步入历史上最好的发展时期。

四川地处长江上游，山川纵横的特点，造就了川菜独特的内陆性特征；四川历史上的社会变动和人口迁移，又使川菜拥有和其他内陆地区不一样的开放性。具体而言，川菜的主要特点是取材广泛、调味多变、技法多样、品类繁多。

第一是取材广泛。四川素有"天府之国"之称，其境内沃野千里，江河纵横。优越的自然条件，为川菜体系的建立提供了丰富而优质的烹饪原料。淡水鱼中的佳品有江团、雅鱼、石爬鱼、鲶鱼、鳙鱼、鲫鱼等；蔬菜中的名特产原料有葵菜、豌豆尖、莴笋、韭黄、红油菜薹、青菜头、藠头、红心萝卜、甜椒等。干杂品如通江、万源的银耳，宜宾、乐山、凉山的竹笋，青川、广元的黑木耳，宜宾、达县的香菇，渠县、南充的黄花菜及多种菌类植物等，均堪称佼佼者。就连生长在田边地头、深山河谷中的野蔬之品也成为川菜的好原料，如石耳、地耳、苕菜、侧耳根、马齿苋等。值得注意的是，川菜取材广泛，但不以食材的古怪和稀缺为号召力，而是以普通、绿色、健康为选材的基本原则，这一点在当今社会尤为值得赞赏和提倡。

第二是调味多变。川菜的基本味为麻、辣、甜、咸、酸、苦，在此基础上，又可调配变化为多种复合味型。新中国成立之前，四川历史上大规模的人口迁移共有六次，对四川的社会生活产生了重大影响。人口迁移使各地区和各民族的人在四川共同生活，既把他们原有的饮食习俗、烹调技艺带进了四川，又受到四川原住居民饮食习俗的影响，互相交流，口味融合，形成并发展为四川地区动态、丰富的口味。同时，讲究饮食滋味的四川人十分注意培育优良的种植调味品和生产高质量的酿造调味品，自贡井盐、内江白糖、阆中保宁醋、中坝酱油、郫县豆瓣、汉源花椒、宜宾芽菜、南充冬菜、成都二荆条辣椒等，这些质地优良的

调味品，为川菜烹饪的调配变化提供了良好的物质基础。外地人谈到川菜，常常认为川菜的风味特点就是麻辣，其实这个看法有失偏颇。川菜真正的风味特色是清鲜醇浓并重，善用麻辣。当今川菜常用的复合味型已达27种，居全国之首，人们赞美川菜是"一菜一格，百菜百味"不无道理。需要强调的是，现代川菜虽然不含辣的菜肴占有很大部分，但是如果没有辣椒，川菜的个性就会大打折扣，而辣椒也因为附丽于调味多变的川菜而更显魅力，才在全国更加流行。

第三是技法多样。川菜的烹调方法极富变化，火候运用极为讲究。川菜的烹饪技法大类有30种，每一类又下分若干种（如炒法类，又再分生炒、熟炒、小炒、软炒；蒸法类，又再分清蒸、旱蒸、粉蒸等等），最终形成数十种常用烹饪技法。川味菜式就是运用多种烹调方法烹制出来的。可以说，每一种技法在烹制川菜时都能各显其妙。其中，最能表现川味特色的技法，当数小炒、干煸、干烧、家常烧。小炒的经典菜肴有鱼香肉丝、宫保鸡丁等，干煸的代表菜肴有干煸牛肉丝、干煸四季豆等，常见干烧的菜肴则有干烧岩鲤、干烧鲫鱼等，家常烧的菜肴则有家常海参、大蒜鲢鱼等。

第四是品类繁多。现代川菜主要由菜肴、面点小吃、火锅三大类型组成，不同类型各具风格特色，又互相渗透配合，形成一个完整的体系，对各地、各阶层的食客有着广泛的适应性。据不完全统计，在20世纪末，川菜的品种已不低于5000种。21世纪以来，随着川菜行业的发展，菜品的创新速度越来越快，新菜源源不断地涌现，因此其数量更加可观。现代川菜品种众多，从制作精细程度与消费层次相结合的角度来划分，有制作精细、适合高档消费的精品川菜，也有制作相对粗犷、适合中低档消费的大众川菜。从技术特点与形成历史相结合的角度来划分，有正宗川菜、传统川菜，也有创新川菜、现代川菜。从技术规范来划分，有学院派川菜与江湖派川菜等类型。而从所在地域来划分，有本土川菜与海派川菜，其中本土川菜又可划分为四川地方菜、重庆地方菜、内江地方菜等等。这些划分和名称，足以表明川菜品种的多样性。

"食在中国，味在四川"。飘香的川菜让世人对物华天宝、人杰地灵的四川多了一份亲近，多了一份喜爱，川菜已经成为四川的一张靓丽名片，成为四川人民友好交往的使者和桥梁。21世纪以来，随着中国社会经济的高速发展，传统的川菜烹饪艺术逐渐与现代的经营、生产方式完美结合，川菜一定会更好更快地走向世界、走向未来。

OVERVIEW of SICHUAN CUISINE

Chinese food is renowned throughout the world, and Sichuan cuisine, as one of the most distinguished and applauded branches of the Chinese culinary culture, enjoys enormous popularity among the Chinese. *"While China is an ideal place to feed your appetite, Sichuan is the best place to satisfy your taste buds."* – This saying vividly depicts the significance of Sichuan cooking.

Sichuan cuisine boasts a long history. It originated from the ancient states Shu and Ba, and continued its development during the Shang, Zhou and Qin dynasties. By the Han and Jin dynasties, distinctive features had been seen in Sichuan cooking, which then embarked on the fastest growth during the Sui, Tang and Wudai dynasties. The Song Dynasty witnessed its entry into the catering market of Beijing, and Ming Dynasty was a time when Sichuan cooking developed steadily. The adoption of chili peppers as one of the seasonings in the Mid-Qing Dynasty further advanced Sichuan cuisine without changing its traditional emphasis on taste and pungency. Toward the end of the Qing Dynasty, Sichuan cuisine developed into a unique school in terms of flavors. Since the founding of New China, especially under the policy of Reform and Opening-up, Sichuan cuisine, like any other school of Chinese cooking, has flourished by being innovative and learning from others.

The location of Sichuan on the upper reaches of the Changjiang River has endowed Sichuan cuisine with typical inland features while the social and population changes in Sichuan history have contributed to its openness compared with any other inland area. Specifically speaking, the main characteristics of Sichuan cuisine are many types of ingredients, varied ways of seasoning, a wide range of cooking techniques and numbers of categories.

Firstly, Sichuan cuisine has a variety of ingredients available. Sichuan, since ancient times, has been referred to as "Land of Abundance", where there are fertile fields and numerous rivers. These provide a plentiful supply of high-quality cooking ingredients. Among them the famous ones are Longsnout catfish, Ya fish, Euchiloglanis davidi, catfish, bighead carp and crucian carp among the fresh-water fish and cluster mallow, pea vine sprouts, asparagus lettuce, leeks, rape, mustard plants, radish and bell peppers among the vegetables. The popular dried ingredients include snow funguses from Tongjiang and Wanyuan, bamboo shoots from Yibin, Leshan and Liangshan, Jew's ear funguses from Qingchuan and Guangyuan, shiitake mushrooms from Yibin and Daxian, and daylily from Quxian and Nanchong. The wild plants that grow at the edge of the cultivated fields, in the deep valleys and on the remote mountains are also culinary ingredients, such as stone funguses, nostoc funguses, milk vetch, heartleaf and purslane. What we should keep in mind is that Sichuan cuisine, while taking advantage of a vast range of cooking materials, focuses not on the rare and weird ingredients but on common, green and healthy ones, which is applauded in the modern society.

Secondly, Sichuan cuisine has various ways to season. The basic flavors of the Sichuan cuisine are numbing, hot, sweet, salty, sour and bitter. The combination of the basic flavors form compound flavors. Before New China was founded, there had been six waves of migration in Sichuan history, which resulted in enormous social changes. People from different regions and ethnic groups had to live together, and they brought with them their distinct dietary habits and cooking techniques. These habits and techniques merged and interacted with those of Sichuan, and finally led to the diverse and dynamic tastes in Sichuan cooking. Meanwhile, the Sichuanese have been striving to cultivate natural spices and produce high-quality seasonings, such as the well salt of Zigong, white sugar of Neijiang, Baoning vinegar of Langzhong, soy sauce of Zhongba, chili bean paste of Pixian, fermented soy beans of Yongchuan, Sichuan pepper of Hanyuan, yacai of Yibin, dongcai of Nanchong, and erjingtiao chili peppers of Chengdu, etc. These condiments are the basis of the Sichuan cooking and its various ways of seasoning. Sichuan cuisine boasts the largest number of compound tastes: 27 in all. What people appreciate of the Sichuan dishes is that every dish has its own style and a hundred dishes have 100 different tastes. One thing worth noting is that, despite the fact that most of the modern Sichuan dishes are not flavored by chili peppers, the distinctiveness of Sichuan cuisine will be greatly reduced without chili peppers and that chili peppers will not be so popular in China without being associated with the many-flavored Sichuan dishes.

Thirdly, Sichuan cuisine has a number of cooking techniques. Sichuan cooking includes a number of cooking techniques and attaches great importance to the use of flames. Broadly speaking, there are 30 cooking ways. The number amounts to over 50 if the sub-branches are counted in. Most Sichuan dishes are prepared with more than one way of cooking, each performing its unique function. The most peculiar cooking method of Sichuan cuisine are sautee, dry-fry, dry-braise and home-style-braise. The courses Fish-Flavor Pork Slivers, Gongbao Diced Chicken are typical of the cooking method—Sautee; Dry-Fried Beef Slivers and Dry-Fried French Beans fully demonstrate the characteristics of dry-frying; Dry-Braised Crucian Carp and Dry-Braised Carp are the ones that mainly employ the technique dry-braising; Home-Style Sea Cucumber and Bighead Carp with Garlic are both of home style.

Fourthly, there are a number of varieties in Sichuan cooking. Modern Sichuan cuisine mainly consists of three categories, namely, dishes, snacks and hot pot, which influence each other while retaining their own properties and catering for different tastes. It is estimated that there were more than 5,000 Sichuan dishes at the end of the 20th century. The 21st century has seen an even greater number of dishes with new recipes invented at a remarkable speed. As far as preparation complexity and target customers are concerned, they range from exquisite and expensive dishes to common ones with reasonable prices. In terms of technical features and history, there are orthodox dishes, traditional dishes, innovative dishes and modern dishes. The division between academic school and vocational school is based on the technical regulations. According to locality, Sichuan dishes fall into overseas ones and local ones, and the latter are further divided into Sichuan dishes, Chongqing dishes, and Neijiang dishes. The various ways of classification reveal the great varieties of the dishes.

"While China is an ideal place to feed your appetite, Sichuan is the best place to satisfy your taste buds." People's love for this beautiful land has been doubled by the savory Sichuan dishes, which are now a bridge to facilitate Sichuan's communication with the other parts of the world. With China's rapid social and economic development in the 21st century and Sichuan cuisine's combination with modern production and management systems, the dishes of Sichuan are bound to act an even more influential part around the world in the future.

CONTENTS 目录

第一篇 | Appetizer 开胃菜

五香牦牛肉	Five-Spice-Flavored Yak Beef	003
四味鲍鱼	Four-Flavor Abalone	004
葱酥鱼	Crispy Scallion-Flavored Fish	005
怪味鸡丝	Multi-Flavored Chicken Slivers	007
椒麻鸡	Jiaoma-Flavor Chicken	008
白宰鸡	Baizai Chicken	009
花椒鸡丁	Sichuan-Pepper-Flavored Chicken	011
泡椒凤爪	Pickled-Chili-Flavored Chicken Feet	012
钵钵鸡	Bobo Chicken	013
腊味拼盘	Cured Meat Platter	014
卤水拼盘	Assorted Meat Stewed in Sichuan-Style Broth	015
太白酱肉	Taibai Flour-Paste-Flavored Pork	017
蒜泥白肉	Pork in Garlic Sauce	018
酱猪手	Flour-Paste-Flavored Pork Feet	021
糖醋排骨	Sweet-and-Sour Spareribs	022
烟熏排骨	Smoked Spareribs	025
芥末肚丝	Tripe Slivers in Mustard Sauce	026
红油耳片	Pork Ear Slices in Chili Oil	027
陈皮兔丁	Tangerine Peel-Flavored Rabbit Dices	028
夫妻肺片	Fuqi Feipian (Sliced Beef & Offal in Chili Sauce)	029
麻辣牛肉干	Mala Dried Beef Strips	031
灯影牛肉	Translucent Beef Slices	032
双味蘸水兔	Rabbit with Double-Flavor Dipping Sauces	034
四川泡菜	Sichuan Pickles	037
老妈兔头	Laoma Rabbit Heads	038
侧耳根拌蚕豆	Sichuan-Style Heartleaf and Broad Bean Salad	039
灯影苕片	Translucent Sweet Potato Chips	041
口口脆	Crunchy Auparagus Lettuce	042
酸辣蕨粉	Hot-and-Sour Fern Root Noodles	043
怪味花仁	Multi-Flavored Peanuts	044

姜汁豇豆	Asparagus Beans in Ginger Sauce	046
泡椒双耳	Black and White Chili-Pickle-Flavored Funguses	047
椒麻桃仁	Jiaoma-Flavor Walnuts	049
荞面鸡丝	Buckwheat Noodles with Shredded Chicken	050
麻酱凤尾	Asparagus Lettuce with Sesame Paste	051
鱼香豌豆	Peas in Fish-Flavor Sauce	052

第二篇 热 菜 Hot Dishes

◆ 海鲜类 Seafood

红烧鲍鱼	Red-Braised Abalone	056
宫保龙虾球	Gongbao Lobster Balls	059
鱼香龙虾	Lobster in Fish-Flavor Sauce	060
干烧大虾	Dry-Braised Prawns	063
翡翠虾仁	Shrimps with Jade-Colored Broad Beans	064
盆盆虾	Penpen Prawns (Spicy Prawns in a Basin)	065
干烧辽参	Dry-Braised Liaoning Sea Cucumber	067
家常海参	Home-Style Sea Cucumber	068
酸辣海参	Hot-and-Sour Sea Cucumber	069
白汁鱼肚卷	Fish Maw Rolls in Milky Sauce	070
椒汁多宝鱼	Turbot in Pepper-Flavored Sauce	071
菠饺鱼肚	Spinach-Flavored Dumplings with Fish Maw	072
家常鱿鱼	Home-style Squid	074
家常鱼唇	Home-style Fish Snouts	075
干煸鱿鱼丝	Dry-Fried Squid Slivers	076
泡椒墨鱼仔	Pickled-Chili-Flavored Tiny Cuttlefish	077
荔枝鱿鱼卷	Lichi-Flavor Squid Rolls	079
香辣蟹	Hot-and-Spicy Crabs	080
煳辣鲜贝	Hula-Flavor Scallops	083
竹烤银鳕鱼	Roasted Cod on a Bamboo Platter	084
藿香鲈鱼	Ageratum-Flavored Perch	086
双椒石斑鱼	Speckled Hind Fish with Green and Red Peppers	087

◆ 山珍类 Mountain Delicacies

冰糖燕窝	Bird's Nest with Rock Sugar	088
清汤燕菜	Bird's Nest in Consomme	089
一品牦牛掌	Deluxe Yak Paws	090
竹荪鸽蛋	Pigeon Eggs with Veiled Lady Mushrooms	093

◆ 河鲜类 River Delicacies

清蒸百花江团	Steamed Longsnout Catfish Surrounded by Flowers	095
红烧裙边	Red-Braised Shell Rims of Chinese Turtle	096
土豆烧甲鱼	Braised Chinese Turtle with Potatoes	097
川式烤鳗鱼	Sichuan-Style Barbecued Eel	099
豆瓣鱼	Fish in Chili Bean Sauce	100
砂锅雅鱼	Ya Fish Casserole	101
香辣黄蜡丁	Hot-and-Spicy Yellow Catfish	103
麻辣小龙虾	Mala Crayfish	104
泡椒牛蛙	Pickled-Chili-Flavored Bullfrog	105
开门红	Good-Luck Fish Head	106
石锅牛蛙	Bullfrog in a Stone Pot	107
川味烤鱼	Sichuan-Flavor Barbecued Fish	109
干烧鱼	Dry-Braised Fish	110
糖醋脆皮鱼	Crispy Sweet-and-Sour Fish	111
芹黄熘鱼丝	Stir-Fried Fish Slivers with Celery	113
酸菜鱼	Fish with Pickled Mustard	114
软烧仔鲶	Braised Catfish	115
香辣沸腾鱼	Hot-and-Spicy Sizzling Fish	116
鳝段粉丝	Paddy Eels with Pea Vermicelli	119
干煸鳝丝	Dry-Fried Paddy Eel Slivers	120
大蒜烧鳝鱼	Braised Paddy Eels with Garlic	122
香辣炝泥鳅	Hot-and-Spicy Loach	123

◆ 禽肉类 Poultry

宫保鸡丁	Gongbao Diced Chicken	125
太白鸡	Taibai Chicken	126
鸡米杂粮配窝窝头	Chopped Chicken with Steamed Corn Buns	127
野生菌煨乌鸡	Stewed Silkie Chicken with Wild Mushrooms	128
松茸炖土鸡	Stewed Free-range Chicken with Matsutake	129
芙蓉鸡片	Hibiscus-like Chicken	130

辣子鸡丁	Diced Chicken with Pickled Chilies	131
鸡豆花	Chicken Curd	133
鸡蒙葵菜	Cluster Mallow Coated with Chicken Mince	134
白果炖鸡	Stewed Chicken with Gingko Nuts	136
黄焖鸡	Golden Chicken Stew	137
鱼香八块鸡	Chicken Chunks in Fish-Flavor Sauce	138
雪花鸡淖	Snowy Chicken	141
香辣掌中宝	Hot-and-Spicy Chicken Feet Pad	142
青椒鸡杂	Chicken Hotchpotch with Green Peppers	145
虫草鸭子	Steamed Duck with Caterpillar Fungus	146
樟茶鸭	Tea-Smoked Duck	149
甜皮鸭	Crispy Sweet-Skinned Duck	150
姜爆鸭丝	Quick-Fried Duck Slivers with Ginger	151
酱爆鸭舌	Quick-Fried Duck Tongues with Fermented Flour Paste	153
香辣鸭唇	Hot-and-Spicy Duck Jaws	154
香酥鸭子	Crispy Duck	157
鸡㞢烩鸭腰	Braised Duck Kidneys with Collybia Mushrooms	158
天麻乳鸽	Pigeon Stew with Gastrodia Tuber	161
鱼香虎皮鸽蛋	Tiger-skin Pigeon Eggs in Fish-Flavor Sauce	162

◆ 畜肉类 Meat

回锅肉	Twice-Cooked Pork	164
盐煎肉	Stir-Fried Pork with Leeks	166
东坡肘子	Dongpo Pork Knuckle	167
红烧肉	Red-Braised Pork Belly	169
坛子肉	Stewed Meat in an Earthen Pot	170
鱼香肉丝	Pork Slivers in Fish-Flavor Sauce	173
酱肉丝	Stir-Fried Pork Slivers with Fermented Flour Paste	174
青椒肉丝	Pork Slivers with Green Peppers	177
锅巴肉片	Sliced Pork with Sizzling Rice Crust	178
粉蒸肉	Steamed Pork Belly with Rice Flour	181
糖醋里脊	Sweet-and-Sour Pork Tenderloin	182
咸烧白	Steamed Pork with Salty Stuffing	185
甜烧白	Steamed Pork with Sweet Stuffing	186
火爆双脆	Crispy Quick-Fried Pork Tripe and Chicken Gizzards	187

火爆腰花	Quick-Fried Pork Kidneys	189
萝卜连锅汤	Pork Soup with Radish	191
香辣猪蹄	Hot-and-Spicy Pork Feet	193
雪豆蹄花	Pork Feet Stew with White Haricot Beans	194
葱烧蹄筋	Braised Pork Tendon with Scallion	195
川椒牛仔骨	Sichuan-Style Pepper-Flavored Beef Spareribs	197
红烧牛头方	Red-Braised Water Buffalo Scalp	198
干煸牛肉丝	Dry-Fried Beef Slivers	199
水煮牛肉	Boiled Beef in Chili Sauce	201
小笼蒸牛肉	Steamed Beef in a Small Bamboo Steamer	203
竹笋烧牛肉	Braised Beef with Bamboo Shoots	205
香辣肥牛	Hot-and-Spicy Beef	206
藤椒肥牛	Beef with Green Sichuan Pepper	209
鲜椒仔兔	Rabbits with Chili Peppers	210

◆ 素菜类 Vegetables

麻婆豆腐	Mapo Tofu	212
家常豆腐	Home-Style Tofu	215
口袋豆腐	Pocket Tofu	216
过江豆花	Silken Tofu with Dipping Sauce	217
砂锅豆腐	Tofu Casserole	219
毛血旺	Duck Blood Curd in Chili Sauce	220
开水白菜	Napa Cabbage in Consomme	223
鱼香茄饼	Eggplant Fritters in Fish-Flavor Sauce	224
臊子蒸蛋	Steamed Egg with Topping	226
白油苦笋	Stir-Fried Bitter Bamboo Shoots	227
酱烧冬笋	Braised Winter Bamboo Shoots with Fermented Flour Paste	229
干锅茶树菇	Black Poplar Mushrooms in a Small Wok	230
干贝菜心	Napa Cabbage with Dried Scallops	233
干煸四季豆	Dry-Fried French Beans	234
蚕豆泥	Mashed Broad Beans	235
川贝酿雪梨	Pear Stuffed with Fritillaria Cirrhosa	236
金沙玉米	Golden-Sand Corn (Fried Corn with Egg Yolk)	239
番茄蛋花汤	Tomato and Egg Soup	240
绿豆南瓜汤	Pumpkin Soup with Mung Beans	243

第三篇 火锅 — Hot Pot

毛肚火锅	Beef Tripe Hot Pot	246
鸳鸯火锅	Double-Flavor Hot Pot	249
羊肉汤锅	Mutton Soup Hot Pot	250
串串香	Chuan Chuan Xiang Hot Pot	253
冷锅鱼	Fish in Cold Pot	254
干锅鸡	Sauteed Chicken in a Small Wok	257

第四篇 面点小吃 — Snacks

担担面	Dandan Noodles	260
钟水饺	Zhong's Dumplings	262
川北凉粉	Northern-Sichuan-Style Pea Jelly	263
龙抄手	Long Wonton	264
牛肉焦饼	Crispy Pancakes with Beef Stuffing	267
小笼包子	Steamed Buns in Small Bamboo Steamers	269
叶儿粑	Leave-Wrapped Rice Dumplings	270
珍珠圆子	Pearly Tangyuan	273
蛋烘糕	Dan Hong Gao (Sichuan-Style Stuffed Pancakes)	274
黄粑	Brown Rice Cake Wrapped in Leaves	277
鸡丝凉面	Cold Noodles with Shredded Chicken	278
军屯锅盔	Juntun Pancakes	280
三大炮	Three Cannonshots (Sweet Rice Buns)	281
赖汤圆	Lai's Tangyuan (Sweet Rice Dumplings)	283
铺盖面	Sheet Pasta with Topping	284
红烧牛肉面	Noodles with Red-Braised Beef Topping	286
酸辣粉	Hot-and-Sour Sweet Potato Noodles	287
一品锅贴	Deluxe Fried Dumplings	289

附录 Appendix

川菜特色调味品	Featured Seasonings	292
川菜烹饪术语	Terms	296

Sichuan (China) Cuisine in Both Chinese and English

川菜
(中英文标准对照版)

开胃菜
Appetizer

第一篇

 色泽棕红，牛肉香浓，回味悠长
dark brown color; moreish and savoury beef; lingering aroma

[原 料]
牦牛肉500克

[调料A]
食盐3克 料酒20克 姜10克 葱段15克

[调料B]
白卤水2000克 料酒20克 姜10克 葱段20克

[制 作]
1. 牦牛肉洗净后切成两块，用调料A码味4～6小时。
2. 将调料B放入陶瓷锅中，再放入码好味的牦牛肉块，用小火煮2～3小时，至牦牛肉软熟时捞出，冷后切成条即成。

Ingredients
500g yak beef

Seasonings A
3g salt, 20g Shaoxing cooking wine, 10g ginger, 15g scallion (cut into sections)

Seasonings B
2000g Sichuan-style broth, 20g Shaoxing cooking wine, 10g ginger, 20g scallion (cut into sections)

Preparation
1. Wash the yak beef thoroughly, halve, and marinate in Seasonings A for 4 to 6 hours.
2. Pour Seasonings B into an earthen pot, add the beef and simmer for 2 to 3 hours till the beef becomes soft and cooked through. Remove and cut into strips when cool.

五香牦牛肉
Five-Spice-Flavored Yak Beef

四味鲍鱼
Four-Flavor Abalone

色淡雅，质鲜嫩，形美观

beautifully-arranged ingredients; natural color; delicate and tender abalone

[原 料]

罐装鲍鱼400克 粉皮150克 菜心100克

[调 料]

芥末味碟、怪味味碟、椒麻味碟、麻酱味碟各1碟 味精1克 葱段10克 姜片10克 料酒10克 鲜汤500克

[制 作]

1. 锅中放鲜汤、姜片、葱段、料酒、味精烧沸，倒入装有鲍鱼的汤碗中喂味15分钟，取出鲍鱼晾凉后逐个片成薄片。
2. 菜心烫至刚熟捞出，晾凉后摆入盘中垫底；粉皮切成菱形片，放在菜心中间；鲍鱼片摆放在粉皮上，配四个味碟上桌。

Ingredients

400g canned abalone, 150g steamed mung bean jelly, 100g tender leafy vegetables

Seasonings

mustard-flavored sauce served on a saucer, multi-flavored sauce served on a saucer, jiaoma-flavored sauce served on a saucer, sesame-paste-flavored sauce served on a saucer, 1g MSG, 10g scallion (cut into sections), 10g ginger (sliced), 10g Shaoxing cooking wine, 500g everyday stock

Preparation

1. Add the stock, ginger, scallion, Shaoxing cooking wine and MSG to a wok, bring to a boil and transfer to a bowl containing the abalone. Marinate the abalone for 15 minites, remove, cool and cut into thin slices.
2. Blanch the vegetables till al dente, cool and lay neatly on a serving dish. Cut the bean jelly into diamonds and put onto the vegetables. Lay the abalone slices on the bean jelly, and serve with the four saucers of dipping sauce.

色泽棕黄，肉嫩骨酥，咸鲜味醇，葱香浓郁

deep brown color; tender fish; crispy bones; rich tastes enhanced by scallion

葱酥鱼
Crispy Scallion-Flavored Fish

[原 料]

鲫鱼500克 水发玉兰片50克 水发香菇50克 食用油1000克（约耗80克）

[调 料]

泡辣椒段30克 葱白段150克 姜片10克 食盐4克 料酒30克 醪糟汁50克 糖色15克 味精1克 醋5克 鲜汤300克 芝麻油5克

[制 作]

1. 鲫鱼初加工后在鱼身两侧各剞3刀，用姜片、葱段、食盐、料酒码味10分钟；玉兰片、香菇片成片。
2. 锅中放油烧至200℃，将鱼入锅炸至色浅黄时捞出。
3. 锅内留油少许，放入泡辣椒段、葱白段炒香，掺入鲜汤，放入食盐、料酒、醋、糖色、醪糟汁、玉兰片、香菇、鲫鱼，用小火收制。待鱼软熟入味时捞出，锅中的汁用旺火收至将干时，放入味精、芝麻油，连汁一起倒在鱼上，待鱼冷后装盘成菜。

Ingredients

500g crucian carp, 50g water-soaked bamboo shoot slices, 50g water-soaked shiitake mushrooms, 1000g cooking oil for deep-frying

Seasonings

30g pickled chilies (cut into sections), 150g scallion (white part only, cut into sections), 10g ginger (sliced), 4g salt, 30g Shaoxing cooking wine, 50g fermented glutinous rice wine, 15g caramel colour, 1g MSG, 5g vinegar, 300g everyday stock, 5g sesame oil

Preparation

1. Kill the crucian carp, and clean thoroughly. Make three cuts into each side of each fish, and marinate in the mixture of ginger, salt, scallion and Shaoxing cooking wine for ten minutes. Slice the shiitake mushrooms.
2. Heat oil in a wok to 200℃, and deep-fry the fish till brownish.
3. Drain off most of the oil, and stir-fry in the remaining oil the pickled chilies and scallion to bring out their aroma. Add the stock, salt, Shaoxing cooking wine, vinegar, caramel color, fermented glutinous rice wine, bamboo shoot slices, shiitake mushrooms and crucian carp. Simmer over a low flame till the carp is soft and fully cooked. Remove the carp and continue to heat the wok over a high flame to reduce the sauce. Add the MSG and sesame oil, and then pour the sauce over the fish. Wait till the fish is cool, and transfer to a serving dish.

色泽红亮，咸、甜、麻、辣、鲜、香、酸诸味兼之
lustrous appealing color; a mixture of flavors

怪味鸡丝
Multi-Flavored Chicken Slivers

[原　料]
熟公鸡肉150克　葱丝20克

[调　料]
芝麻酱15克　辣椒油50克　食盐2克　白糖15克　酱油20克　醋15克　味精3克　花椒粉3克　芝麻油5克　熟芝麻6克

[制　作]
1. 熟公鸡肉切成丝，放在垫有葱丝的盘内。
2. 将芝麻酱、辣椒油、食盐、白糖、酱油、醋、味精、花椒粉、芝麻油调匀，再放入熟芝麻调匀成怪味汁，食用时淋在鸡丝上即成。

Ingredients
150g precooked rooster meat, 20g scallion (cut into slivers)

Seasonings
15g sesame paste, 50g oil-infused chili flakes, 2g salt, 15g sugar, 20g soy sauce, 15g vinegar, 3g MSG, 3g ground roasted Sichuan pepper, 5g sesame oil, 6g roasted sesame seeds

Preparation
1. Cut the chicken into slivers. Lay the scallion slivers on a serving dish and arrange the chicken slivers on top.
2. Mix the sesame paste, oil-infused chili flakes, salt, sugar, soy sauce, vinegar, MSG, Sichuan pepper, sesame oil and sesame seeds to make seasoning sauce. Drizzle the sauce over the chicken slivers.

Jiaoma-Flavor Chicken

色泽红亮，麻辣味厚，有浓厚的葱香和花椒香味

bright reddish color; pungent, tingling and enticing taste; fragrance from scallion and Sichuan pepper

[原　料]
熟鸡肉200克　葱丁20克

[调　料]
食盐2克　酱油10克　椒麻糊10克　味精1克　花椒油5克　辣椒油50克　芝麻油5克

[制　作]
1．熟鸡肉斩成约3厘米宽的长形块。
2．调料调匀成椒麻味汁。
3．鸡块、葱丁与椒麻味汁拌匀，装入盘中即成。

Ingredients
200g precooked chicken, 20g scallion (finely chopped)

Seasonings
2g salt, 10g soy sauce, 10g Jiaoma paste, 1g MSG, 5g Sichuan pepper oil, 50g chili oil, 5g sesame oil

Preparation
1. Chop the precooked chicken into 3cm³ cubes.
2. Mix the seasonings to make jiaoma-flavoar sauce.
3. Mix chicken cubes, chopped scallion and jiaoma-flavor sauce, blend well and transfer to a serving dish.

白宰鸡
Baizai Chicken

肉质鲜嫩，入味充分，麻辣鲜香

tender meat; spicy, zingy and aromatic taste

[原 料]

土公鸡500克 葱丁50克

[调 料]

豆瓣卤汁30克 辣椒油150克 花椒粉5克
白糖25克 食盐5克 味精3克 芝麻油10克

[制 作]

1. 土公鸡清理干净后放入热水锅中，用小火煮熟后取出。
2. 用豆瓣卤汁、辣椒油、花椒粉、白糖、食盐、味精、芝麻油调成味汁。
3. 待鸡完全冷透后，将鸡肉拍松，剁成块，装入垫有葱丁的大盘中，浇入调味汁即可。

Ingredients

500g free-range rooster, 50g scallion (chopped)

Seasonings

30g chili-bean-paste-flavored sauce, 150g chili oil, 5g ground roasted Sichuan pepper, 25g sugar, 5g salt, 3g MSG, 10g sesame oil

Preparation

1. Wash the rooster thoroughly, transfer to a pot containing hot water, bring to a boil, and then simmer over a low flame till cooked through. Remove the rooster from the water and set aside to cool.
2. Mix together the chili-bean-paste-flavored sauce, chili oil, ground roasted Sichuan pepper, sugar, salt, MSG and sesame oil to make the seasoning sauce.
3. Pat the cooled chicken to loosen its fibres, and then cut it into chunks. Transfer the chicken chunks to a platter containing chopped scallion, and pour over the seasoning sauce.

典故

白宰鸡，是四川乐山地区著名菜肴，又称凉拌鸡，因其是将公鸡不加任何调味料而直接煮熟晾凉后拍松、切割，待食用时才加调味汁而得名。

Note

Baizai Chicken ("chopped plain chicken" in Chinese), a renowned dish from Leshan, is so called because the rooster is first boiled in plain water till cooked through without any seasonings. The boiled rooster is then patted and cut. The seasoning sauce is not added until the chicken is served.

色泽金红，酥软化渣，麻辣鲜醇
appealing color of golden brown; tender chicken that is crispy on the outside; spicy and pungent taste

花椒鸡丁

Sichuan-Pepper-Flavored Chicken

[原 料]

鸡腿300克 食用油1000克（约耗100克）

[调 料]

干辣椒节20克 花椒5克 姜片10克 葱段15克 料酒15克 白糖5克 食盐4克 味精2克 芝麻油10克 糖色20克 鲜汤200克

[制 作]

1. 鸡腿去骨，斩成2.5厘米见方的丁，加入食盐、料酒、姜片、葱段码味15分钟。
2. 锅中放油烧至150℃，放入鸡丁炸去表面水分后捞出，待油温升至180℃时再入油中炸至表面浅黄、酥香时捞出。
3. 锅中放油烧至120℃，下干辣椒节、花椒炒香，再下鸡丁略炒，加入鲜汤、食盐、白糖、糖色、料酒，用小火收汁，最后下味精、芝麻油，起锅晾冷后装盘即成。

Ingredients

300g chicken legs, 1000g cooking oil for deep-frying

Seasonings

20g dried chilies (chunked), 5g Sichuan pepper, 10g ginger (sliced), 15g scallion (cut into sections), 15g Shaoxing cooking wine, 5g sugar, 4g salt, 2g MSG, 10g sesame oil, 20g caramel color, 20g stock

Preparation

1. De-bone the chicken legs and cut into 2.5cm³ dices. Mix the chicken with salt, Shaoxing cooking wine, ginger and scallion, and marinate for 15 minutes.
2. Heat the oil in a wok to 150℃, deep-fry the chicken dices till their outer sufaces are dry, then remove from the oil. Reheat the oil to 180℃, add the chicken for a second fry till the surface becomes brownish and crispy.
3. Heat some oil in a wok to 120℃, add dried chilies and Sichuan pepper, and then stir-fry till aromatic. Blend in the chicken, sauté, and add the stock, salt, sugar, caramel color and Shaoxing cooking wine. Simmer till the sauce thickens, add MSG and sesame oil, remove from the heat and transfer to a serving dish when the chicken cools.

泡椒凤爪

色泽天然，鲜香清爽

Pickled-Chili-Flavored Chicken Feet

natural color; mild and delectable taste

[原　料]

鸡爪300克　纯净水600克

[调　料]

野山椒100克　花椒5克　姜片10克　葱段20克　料酒20克　大蒜10克　食盐24克　白糖5克　白醋5克　鸡精3克　胡椒粉2克　干辣椒10克

[制　作]

1. 鸡爪洗净后去指尖，对半切开，入沸水里氽水（煮熟）后捞出、洗净；锅中放入清水、姜片、葱段、料酒、花椒、鸡爪，用中小火煮6分钟后捞出，用冷水冲漂至冷。
2. 将纯净水、野山椒、大蒜、食盐、白糖、白醋、鸡精、胡椒粉、干辣椒放入盆中调成专用泡菜水，放入鸡爪后加盖浸泡，入冰箱中冷藏2小时至入味即成。

Ingredients

300g chicken feet, 600g water

Seasonings

100g pickled tabasco peppers, 5g Sichuan pepper, 10g ginger (sliced), 20g scallion (cut into sections), 20g Shaoxing cooking wine, 10g garlic, 24g salt, 5g sugar, 5g vinegar, 3g chicken essence granules, 2g ground white pepper, 10g dried chilies

Preparation

1. Wash thoroughly the chicken feet and remove the claws. Halve each foot, blanch in boiling water, remove and rinse. Mix some water, ginger, scallion, Shaoxing cooking wine, Sichuan pepper and chicken feet in a wok, then boil over a medium-high flame for 6 minutes. Remove and rinse to cool.
2. Add water, pickled tabasco peppers, garlic, salt, sugar, vinegar, chicken essence granules, ground white pepper and dried chilies to a basin, and blend well to make pickled-chili-flavored brine. Blend in the chicken feet, cover, and preserve in a fridge. Marinate for 2 hours so that the chicken feet could fully absorb the flavors.

Bobo Chicken
钵钵鸡

色泽红亮，肉质细嫩，咸鲜麻辣，味厚爽口
bright color; tender chicken; piquant and zingy taste

[原料]
熟土鸡肉200克 酥花仁40克 葱丁30克

[调料A]
食盐1克 酱油10克 白糖5克 芝麻酱10克 鸡精1克 芝麻辣椒油50克 芝麻油5克 花椒粉2克 花椒油3克

[调料B]
熟芝麻15克 香菜2根

[制作]
1. 熟土鸡肉斩成6厘米长的条，摆放在垫有葱丁、酥花仁的陶钵盛器内。
2. 将调料A调匀成麻辣味汁淋在鸡块上，再撒上调料B即成。

Ingredients
200g precooked free-range chicken meat, 40g crispy peanuts (roasted or fried), 30g scallion (finely chopped)

Seasonings A
1g salt, 10g soy sauce, 5g sugar, 10g sesame paste, 1g chicken essence granules, 50g chili oil with sesame seeds, 5g sesame oil, 2g ground roasted Sichuan pepper, 3g Sichuan pepper oil

Seasonings B
15g roasted sesame seeds, 2 fresh coriander plants

Preparation
1. Cut the precooked chicken into 6cm-long strips, and lay them out in a clay bowl with the scallion and peanuts.
2. Mix Seasonings A evenly to make seasoning sauce and pour over the chicken strips. Sprinkle with the sesame seeds and garnish with coriander.

典故
钵钵鸡来源于四川农村，已有上百年的历史，是用瓦罐盛装的凉拌麻辣鸡，因四川民间称瓦罐为"钵钵"而得名。

Note
Bobo chicken, with its origin in the countryside of Sichuan, has a history of over a hundred years. The dish is so called because the clay bowl used to hold the chicken is known among the Sichuanese as "Bobo".

Cured Meat Platter
腊味拼盘

鲜香味美，回味悠长
lingering aroma

[原　料]

青城山老腊肉100克　腊猪舌100克　金银猪肝60克　香肠100克　花仁150克

[调　料]

卤水750克

[制　作]

1. 花仁放入卤水中煮至软熟后捞出，放在盘中垫底。
2. 青城山老腊肉、腊猪舌、金银猪肝、香肠分别煮熟（或蒸熟）后捞出，晾冷后切成片，摆放在花仁上入笼蒸热即成。

Ingredients
100g Qingcheng Mountain Bacon, 100g smoked pork tongue, 60g fat-stuffed pork liver, 100g Sichuanese wind-dried sausage, 150g peanuts

Seasonings
750g Sichuan-style broth

Preparation
1. Boil the peanuts in the broth till soft and cooked through. Remove and lay onto a serving dish.
2. Boil or steam the Qingcheng Mountain bacon, smoked pork tongue, fat-stuffed pork liver and sausage till cooked through. Remove, cool, slice and lay neatly over the peanuts. Transfer to a steamer and steam till hot.

卤味浓厚，鲜香可口
aromatic smell; dainty and appetizing taste

卤水拼盘

Assorted Meat Stewed in Sichuan-Style Broth

[原　料]

牛蜂窝肚200克　猪舌150克　猪心100克　鹅胗100克　豆腐干100克

[调料A]

卤水2000克　食盐20克　生姜20克　葱20克　料酒150克　玫瑰露酒20克　鸡精5克

[调料B]

鸡精1克　芝麻油3克

[制　作]

1. 将牛蜂窝肚、猪舌、猪心、鹅胗洗净，分别余水（煮沸）后放入卤水锅中，加调料A，用小火卤约1小时至熟后捞出切片；豆腐干另用卤水煮10分钟后捞出切片。用豆腐干垫底，牛蜂窝肚、猪舌、猪心、鹅胗整齐地摆放在上面。
2. 取锅中卤水1勺与调料B调匀成味汁，淋在拼盘上，入笼蒸热即成。

Ingredients

200g beef tripe, 150 pork tongue, 100g pork heart, 100g goose gizzards, 100g dried tofu

Seasonings A

2000g Sichuan-style broth, 20g salt, 20g ginger, 20g scallion, 150 Shaoxing cooking wine, 20g rose wine, 5g chicken essence granules

Seasonings B

1g chicken essence granules, 3g sesame oil

Preparation

1. Rinse the beef tripe, pork tongue, pork heart, goose gizzards, blanch and put into a pot containing some broth. Add Seasonings A, simmer the meat for about an hour, remove and slice. Add some broth in another pot, boil the dried tofu for 10 minutes, remove, slice and lay on a serving dish. Place the stewed meat on top of the dried tofu slices.
2. Mix one ladle of the pre-simmered broth with Seasonings B to make seasoning sauce. Drizzle the meat with the sauce, and steam in a steamer till hot.

酱香浓郁，爽口不腻
strong fermented-flour-paste aroma; savoury taste

太白酱肉
Taibai Flour-Paste-Flavored Pork

[原　料]

带皮猪腿肉500克

[调　料]

食盐10克　花椒5克　白酒15克　甜面酱30克　白糖10克　醪糟汁15克　香料粉5克

[制　作]

1. 猪腿肉切成长条，先用食盐、花椒、白酒腌渍7天，再用食盐、甜面酱、醪糟汁、白糖、香料粉调制成的酱料抹匀，挂通风处晾至略干。
2. 用温水洗去生酱肉表面的酱料，入笼蒸熟，切成薄片后装盘即成。

Ingredients
500g pork leg meat with skin attached

Seasonings
10g salt, 5g Sichuan pepper, 15g alcohol, 30g fermented flour paste, 10g sugar, 15g fermented glutinous rice wine, 5g mixed herbal spices

Preparation
1. Cut the pork leg meat into strips, marinate in the mixture of salt, Sichuan pepper and alcohol for 7 days, and smear with the paste made of salt, fermented flour paste, fermented glutinous rice wine, sugar and herbal spices. Air dry the pork in a well-ventilated place.
2. Rinse the pork with warm water to remove the surface coating. Steam, slice and transfer to a serving dish.

色泽红亮，蒜味浓厚，香辣鲜美，肉质软嫩，肥而不腻

bright and lustrous color; pungent taste; fatty but not greasy pork; strong garlic aroma

蒜泥白肉
Pork in Garlic Sauce

[原　料]

带皮猪坐臀肉300克　葱片10克

[调　料]

蒜泥30克　食盐0.5克　味精1克　复制红酱油40克
辣椒油30克　芝麻油5克

[制　作]

1. 猪肉入锅煮至刚熟后捞出，用原汤浸泡约20分钟至温热。
2. 捞出猪肉揾干水分，片成长约10厘米、宽约5厘米的薄片，装入垫有葱片的盘内。
3. 调料入碗拌匀成蒜泥味，浇淋在肉片上即成。

Ingredients

300g pork rump meat, 10g scallion (sliced)

Seasonings

30g garlic (finely chopped), 0.5g salt, 1g MSG, 40g concocted soy sauce, 30g chili oil, 5g sesame oil

Preparation

1. Boil the pork in a pot till just cooked, transfer the pork to a bowl, pour some broth in the pot into the bowl and soak for about 20 minutes until the pork becomes moderately hot.
2. Remove the pork from the broth, drain, cut into 10×5cm thin slices and transfer to a serving dish with sliced scallion spread as the bottom layer.
3. Mix the Seasonings in a bowl to make garlic-flavored sauce and pour over the pork.

蒜泥白肉
Pork in Garlic Sauce

酱猪手

Flour-Paste-Flavored Pork Feet

色红质糯，咸鲜香醇，味浓郁

rich red-brown color; soft and glutinous pork feet; strong and lingering aroma

酱猪手

Flour-Paste-Flavored Pork Feet

[原　料]
猪蹄3个

[调　料]
酱卤水2000克　料酒10克　葱段20克　姜片15克

[制　作]
1. 猪蹄洗净，入沸水锅内煮3分钟后捞出，用清水洗净。
2. 猪蹄放入酱卤水锅中，加料酒、姜片、葱段，大火烧开后改小火焖煮2小时，捞出后切块装盘即成。

Ingredients
3 pork feet

Seasonings
2000g Sichuan-style broth flavored with fermented flour paste, 10g Shaoxing cooking wine, 20g scallion (cut into sections), 15g ginger (sliced)

Preparation
1. Rinse the pork feet, boil in water for 3 minutes, then remove and rinse in cold water.
2. Place the pork feet in a pot containing the broth, add the Shaoxing cooking wine, ginger, and scallion, and bring to a boil over a high flame before simmering for 2 hours over a low flame. Remove and transfer to a serving dish.

糖醋排骨
Sweet-and-Sour Spareribs

色泽红亮，甜酸醇厚
bright brown color; sweet, sour and savoury taste

[原 料]

猪排骨500克 食用油1500克（约耗50克）

[调 料]

姜片5克 葱段10克 花椒2克 食盐3克 白糖100克 醋30克 料酒15克 熟芝麻5克 芝麻油5克 鲜汤300克

[制 作]

1. 猪排骨剁成5厘米长的段，汆水后洗净装入盆中，加食盐、花椒、姜片、葱段、料酒、清水蒸至排骨成熟时取出，再放入油锅中炸成金黄色捞出。
2. 锅内放食用油烧热，下白糖炒至溶化变色，放排骨、鲜汤、食盐、料酒、白糖，用小火收汁，快干时加醋、芝麻油起锅。
3. 晾凉后撒上熟芝麻拌匀，装盘即成。

Ingredients
500g pork spareribs, 1500g cooking oil for deep-frying

Seasonings
5g ginger (sliced), 10g scallion (cut into sections), 2g Sichuan pepper, 3g salt, 100g sugar, 30g vinegar, 15g Shaoxing cooking wine, 5g roasted sesame seeds, 5g sesame oil, 300g stock

Preparation
1. Chop the spareribs into 5cm-long sections. Blanch, wash and then transfer to a bowl. Add salt, Sichuan pepper, ginger, scallion, Shaoxing cooking wine and water. Steam till the spareribs are cooked through, and then deep-fry them till golden brown.
2. Heat oil in a wok, add sugar, stir-fry till the sugar melts and browns. Add the spareribs, stock, salt, Shaoxing cooking wine and simmer over a low flame till the sauce thickens. Add vinegar and sesame oil before the sauce dries, and then remove the wok from the fire.
3. Wait till the spareribs cool, sprinkle with roasted sesame seeds and transfer to a serving dish.

味美肉嫩，暗含烟香
smoky and aromatic smell; dainty and delicate taste

烟熏排骨
Smoked Spareribs

［原　料］
　　猪排骨1000克

［调　料］
　　食盐4克　花椒5克　料酒20克　胡椒粉1克　醪糟汁20克　八角1个　芝麻油5克

［制　作］
1．调料调匀后均匀地抹在排骨上腌渍1天。
2．用樟树叶、稻草、松柏枝作熏料，把排骨放入熏炉熏至排骨呈浅棕红色、烟味浓郁时出炉，装入大盘，入笼蒸1小时后取出，斩成5厘米长的段，淋上芝麻油即成。

Ingredients
1000g pork spareribs

Seasonings
4g salt, 5g Sichuan pepper, 20g Shaoxing cooking wine, 1g ground white pepper, 20g fermented glutinous rice wine, 1 star anise, 5g sesame oil

Preparation
1. Mix the Seasonings, smear the mixture on the spareribs, and marinate for a day.
2. Smoke the spareribs with fuels made up of the camphor laurel leaves, straws, pine and cypress branches till browned and smoky. Transfer to a large plate and steam in a steamer for 1 hour. Remove from the steamer, chop into 5cm-long sections and drizzle with sesame oil.

芥末肚丝

Tripe Slivers in Mustard Sauce

质地软韧，咸酸鲜辣冲诸味谐调

soft but tenacious tripe; salty, sour and appetizing taste; burning sensation from mustard paste

[原　料]

熟猪肚150克　青笋10克

[调　料]

芥末膏10克　食盐2克　味精1克　酱油6克　醋10克　白糖5克　芝麻油5克　冷鲜汤20克

[制　作]

1. 青笋切丝，入盘垫底；熟猪肚切丝，放在青笋丝上。
2. 芥末膏、食盐、酱油、醋、味精、白糖、芝麻油、冷鲜汤调成芥末味汁，浇淋在肚丝上即成。

Ingredients

150g pre-cooked pork tripe, 10g asparagus lettuce

Seasonings

10g mustard paste, 2g salt, 1g MSG, 6g soy sauce, 10g vinegar, 5g sugar, 5g sesame oil, 20g everyday stock

Preparation

1. Cut the asparagus lettuce into slivers and transfer to a serving dish. Cut the tripe into slivers and lay neatly over the asparagus lettuce.
2. Mix the mustard paste, salt, soy sauce, vinegar, MSG, sugar, sesame oil and stock to make seasoning sauce. Pour the sauce over the tripe slivers.

形大片薄,质地脆嫩,香辣爽口
cripy and tender pork ear; spicy and appetizing taste

Pork Ear Slices in Chili Oil
红油耳片

[原 料]
猪耳朵1只(约300克)

[调 料]
蒜泥10克 辣椒油50克 花椒油10克 食盐3克 酱油15克 醋2克 味精5克 芝麻油5克

[制 作]
1. 猪耳朵刮洗干净,入汤锅用中小火煮40分钟后捞出,用冷水冲漂,冷透后用重物压榨使其平整。
2. 猪耳朵晾干水分,片成长约10厘米、宽约6厘米的薄片装入盘内,将调料调匀成味汁后淋在耳片上即成。

Ingredients
1 pork ear (about 300g)

Seasonings
10g garlic (finely-chopped), 50g chili oil, 10g Sichuan pepper oil, 3g salt, 15g soy sauce, 2g vinegar, 5g MSG, 5g sesame oil

Preparation
1. Clean the ear thoroughly, transfer to a pot and simmer over a medium-low flame for 40 minutes. Rinse in cold water till cool, and then flatten by pressing with heavy objects.
2. Drain the ear and cut into thin slices about 10cm long and 6cm wide. Stack onto a serving dish, mix the seasonings to make the seasoning sauce, and drizzle the sauce over the slices.

Tangerine Peel-Flavored Rabbit Dices

陈皮兔丁

色泽棕红，麻辣咸鲜，陈皮香浓，酥软化渣
deep red-brown color; spicy, piquant and aromatic taste; crispy tangerine peels melting in the diner's mouth

[原　料]

兔肉300克　陈皮15克　食用油1000克（约耗200克）

[调　料]

干辣椒20克　花椒5克　姜10克　葱15克　白糖5克　食盐2克　料酒15克　味精2克　芝麻油10克　糖色15克　鲜汤300克

[制　作]

1. 兔肉切丁，加食盐、料酒、姜、葱码味15分钟；陈皮切片，用清水浸泡待用。
2. 食用油入锅烧至160℃，下兔丁炸至定形捞出，待油温回升到180℃时，将兔丁入锅复炸至色棕黄时捞出。
3. 锅置火上，入干辣椒、花椒、陈皮、兔丁略炒，掺鲜汤，放入食盐、白糖、料酒、糖色烧至汁干，入味精、芝麻油，起锅晾冷后装盘即成。

Ingredients

300g boned rabbit meat, 15g dried tangerine peels, 1000g cooking oil for deep-frying

Seasonings

20g dried chilies, 5g Sichuan pepper, 10g ginger, 15g scallion, 5g sugar, 2g salt, 15g Shaoxing cooking wine, 2g MSG, 10g sesame oil, 15g caramel color, 300g everyday stock

Preparation

1. Dice the rabbit, add salt, Shaoxing cooking wine, ginger and scallion, and marinate for 15 minutes. Slice the tangerine peel and soak in water.
2. Heat oil in a wok to 160℃, deep-fry the rabbit cubes for one or two minutes till they are just cooked, and remove from the oil. Reheat the oil to 180℃, and deep-fry the rabbit cubes for a second time till brown.
3. Heat oil in a wok, add dried chilies, Sichuan pepper, tangerine peels and rabbit dices, sauté and then add some stock, salt, sugar, Shaoxing cooking wine and caramel color. Braise till the liquid has almost completely evaporated before adding the MSG and sesame oil. Remove from the stove, cool and transfer to a serving dish.

色泽红亮，炽软脆爽，麻辣鲜香
bright color; tender meat; spicy appetizing taste

夫妻肺片

Fuqi Feipian (Sliced Beef & Offal in Chili Sauce)

[原　料]

牛肉250克　牛杂（牛心、牛舌、牛头皮）250克

[调　料]

卤水2000克　料酒20克　食盐2克　酱油10克　花椒粉1克　味精1克　辣椒油50克　芝麻粉10克　盐酥花仁20克　葱花10克

[制　作]

1．牛肉、牛杂洗净后分别氽水；盐酥花仁剁碎。
2．将氽水后的牛肉、牛杂放入卤水中，加入料酒煮至软熟，捞出晾冷，切成长约6厘米、宽约3厘米的薄片，混合后装盘。
3．取卤水50克烧沸，盛入碗内，加入食盐、酱油、花椒粉、辣椒油、味精调成味汁，淋在牛肉上，撒上芝麻粉、葱花和碎米盐酥花仁。

典故

20世纪30年代末，成都郭朝华夫妇以经营凉拌肺片为业，他们制作的凉拌肺片深受食者喜爱，被称为"夫妻肺片"。后经厨师改进成目前的样子，但"肺片"一名却沿用至今。

Note

In the late 1930s Chengdu, Guo Chaohua and his wife made a living by selling sliced beef lung salad, which was quite popular and referred to as Fuqi (man-and-wife in Chinese) Feipian (Sliced Lung in Chinese). The name Fuqi Feipian is retained for the improved version of nowadays.

Ingredients

250g beef, 250g beef offal (heart, tongue, head skin)

Seasonings

2000g Sichuan-style broth, 20g Shaoxing cooking wine, 2g salt, 10g soy sauce, 1g ground roasted Sichuan pepper, 1g MSG, 50g chili oil, 10g crushed sesame seeds, 20g crispy salty peanuts, 10g scallion, finely chopped

Preparation

1. Wash the beef and offal thoroughly and blanch in water. Finely Chop the peanuts.
2. Put the blanched beef and offal into the Sichuan-style broth, add the Shaoxing cooking wine and then boil until the beef and offal are soft and cooked through. Remove the beef and offal, cool and then cut into thin slices about 6cm in length and 3cm in width. Mix the slices and then transfer to a serving dish.
3. Bring 50g of the broth to a boil, transfer into a bowl, and add the salt, soy sauce, ground roasted Sichuan pepper and chili oil to make the seasoning sauce. Pour the sauce over the slices and sprinkle with crushed sesame seeds, scallion and chopped peanuts.

色泽红亮，麻辣干香，酥松化渣
bright dark brown color; hot, spicy and aromatic taste; crispy and chewy beef

麻辣牛肉干
Mala Dried Beef Strips

[原　料]
熟牛肉300克 食用油1000克（约耗100克）

[调　料]
辣椒油50克 姜片5克 葱段10克 辣椒粉20克 花椒粉4克 料酒10克 食盐3克 白糖2克 味精1克 芝麻油10克 熟芝麻15克 鲜汤300克

[制　作]
1．熟牛肉切成长约5厘米、粗约1厘米的条，用食盐、姜片、葱段、料酒码味10分钟。
2．锅中放油烧至160℃，入牛肉条炸至色褐红、外表略酥时捞起。
3．锅内留油，加入鲜汤、食盐、味精、白糖、牛肉条，用小火收汁将干时起锅，加入辣椒粉、花椒粉、辣椒油、芝麻油拌匀，晾凉后撒上熟芝麻即成。

Ingredients
300g pre-cooked beef, 1000g cooking oil for deep-frying

Seasonings
50g chili oil, 5g ginger (sliced), 10g scallion (cut into sections), 20g ground chilies, 4g ground roasted Sichuan pepper, 10g Shaoxing cooking wine, 3g salt, 2g sugar, 1g MSG, 10g sesame oil, 15g roasted sesame seeds, 300g everyday stock

Preparation
1. Cut the beef into strips about 5cm long and 1cm thick, add some salt, ginger, scallion and Shaoxing cooking wine, and then marinate for 10 minutes.
2. Heat oil in a wok to 160℃, fry the beef strips till they are dark brown and crispy on the outside.
3. Heat some oil in the wok, add the stock, salt, MSG, sugar and beef strips, and simmer over a low flame to reduce the sauce. Remove from the heat, add ground chilies, ground roasted Sichuan pepper, chili oil and sesame oil, blend well, and then leave to cool. Serve sprinkled with sesame seeds.

片大而薄，红润发亮，麻辣鲜香

broad, thin slices; lustrous dark brown color; spicy and appetizing flavor

灯影牛肉
Translucent Beef Slices

典故

"灯影牛肉"在灯光照射下，能透视出背面的影子，有似皮影戏的效果，故名。

Note

Light can penetrate the thin and translucent beef slices to show the outlines of objects on the other side of the slices, which is rather like shadow-puppet shows.

[原　料]

黄牛肉500克　食用油1500克（约耗80克）

[调　料]

食盐5克　白糖5克　花椒粉5克　辣椒粉10克　五香粉2克　味精2克　芝麻油5克

[制　作]

1. 牛肉滚片成厚薄均匀的大片，撒上食盐，晾干水分。
2. 把晾好后的牛肉平铺在架上，放进炉内用炭火烘干，再上笼蒸约60分钟，趁热切成长约6厘米、宽约4厘米的片。
3. 锅内下食用油烧至120℃，放入牛肉片炸透捞出，加辣椒粉、花椒粉、白糖、味精、五香粉翻簸均匀，晾凉后加芝麻油拌匀即成。

Ingredients

500g beef, 1500g cooking oil for deep-frying

Seasonings

5g salt, 5g sugar, 5g ground roasted Sichuan pepper, 10g ground chilies, 2g five-spice powder, 2g MSG, 5g sesame oil

Preparation

1. Cut the beef into slices, sprinkle with salt and lay out to dry.
2. Scatter the dry beef slices on a grill, and then toast over a charcoal fire to further evaporate the water content. Transfer the toasted slices to a steamer and steam for about 60 minutes. Cut the beef slices into thin slices of 6cm long and 4cm wide while they are still hot.
3. Heat oil in a wok to 120℃ and deep-fry the beef slices. Remove, and add the ground chilies, ground roasted Sichuan pepper, sugar, MSG and five-spice powder. Blend well. Wait till the beef cools, add sesame oil and mix well.

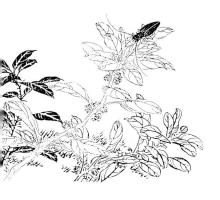

色白质嫩，咸鲜微辣，鲜椒清香
tender and snowy rabbit; spicy and pungent sauce; aroma of fresh chili peppers

双味蘸水兔

Rabbit with Double-Flavor Dipping Sauces

[原　料]
烫皮兔肉750克

[调料A]
八角10克　香叶10克　桂皮10克　姜5克　葱5克　料酒10克
清水2000克

[调料B]
鲜红尖椒末50克　鲜青尖椒末50克　食盐2克　味精1克　芝麻油5克　冷鲜汤20克

[调料C]
香辣酱味碟

操作步骤
1．烫皮兔肉洗净后氽水，再入沸水锅中加入调料A煮熟。
2．待兔肉晾冷后斩成条装盘，配上分别用调料B和调料C制成的味碟，一同上桌即成。

Ingredients
750g rabbit with skin scalded

Seasonings A
10g star anise, 10g bay leaves, 10g cassia bark, 5g ginger, 5g scallion, 10g Shaoxing cooking wine

Seasonings B
50g green chili peppers (finely chopped), 50g red chili peppers (finely chopped), 2g salt, 1g MSG, 5g sesame oil, 20g everyday stock

Seasonings C
chili pepper paste

Preparation
1. Wash the rabbit thoroughly, blanch, and then put into boiling water with Seasonings A. Continue to boil until it is cooked through.
2. Remove the rabbit from the water, cool, chop into long strips and lay onto a serving dish. Serve with dipping sauces made of Seasonings B and Seasonings C respectively.

双味蘸水兔

Rabbit with Double-Flavor Dipping Sauces

色彩丰富，质地脆嫩，咸香可口
natural color; crispy and tender vegetables; salty, fragrant and aromatic taste

四川泡菜
Sichuan Pickles

[原 料]
各种蔬菜（白菜、萝卜、辣椒、芹菜、黄瓜、青笋等）共500克

[调 料]
泡菜盐水100克　凉开水400克　白酒5克　姜片6克　食盐20克　香料3克　花椒3克

[制 作]
1. 去除各种蔬菜的老根、黄叶、粗皮，洗净后切成条块，晾干水分。
2. 取泡菜坛一个，加泡菜盐水、凉开水、白酒、姜片、食盐、香料、花椒，调成新的泡菜盐水，放入蔬菜原料，加盖密封。浸泡5小时后取出即可食用。也可切成丁，加辣椒油、味精、白糖拌匀食用。

Ingredients
500g varieties of vegetables (napa cabbage, radish, chili peppers, celery, cucumber, asparagus lettuce, etc.)

Seasonings
100g pickle brine, 400g cold boiled water, 5g alcohol, 6g ginger (sliced), 20g salt, 3g mixed herbal spices, 3g Sichuan pepper

Preparation
1. Remove any rough roots, withered leaves and rough skin of the vegetables. Rinse, cut into chunks or strips and drain.
2. Mix pickle brine with cold boiled water, alcohol, ginger, salt, herbal spices and Sichuan pepper in an earthen container used exclusively for Sichuan pickles. Add the vegetables. Cover and seal the container, and leave to stand for 5 hours. The pickled vegetables can be served without extra condiments or after being diced and seasoned with chili oil, MSG and sugar.

质嫩入味，麻辣鲜香
tender and savoury meat; aromatic and lingering smell; pungent, numbing and moreish taste

老妈兔头
Laoma Rabbit Heads

[原　料]
鲜兔头1000克

[调　料]
麻辣卤水2000克　姜片20克　葱段20克　食盐15克　料酒30克

[制　作]
1. 鲜兔头冲洗干净，加入姜片、葱段、食盐、料酒拌匀，腌制6小时；用清水洗净后放入沸水锅中氽水，捞出洗净。
2. 卤锅置火上，放入麻辣卤水、兔头，大火烧开后改用小火加热20分钟关火，浸泡2小时，再烧开后浸泡，食用时捞出装盘。

Ingredients
1000g rabbit heads

Seasonings
2000g hot-and-spicy broth of Sichuan style, 20g ginger (sliced), 20g scallion (cut into sections), 15g salt, 30g Shaoxing cooking wine

Preparation
1. Rinse the rabbit heads, add ginger, scallion, salt and Shaoxing cooking wine, blend well and marinate for six hours. Rinse the rabbit heads in water, blanch in boiling water, remove and rinse for a second time.
2. Heat a wok over a flame, add the hot-and-spicy broth and rabbit heads, bring to a boil and then simmer for twenty minutes. Turn off the oven and marinate for 2 hours. Bring to a boil for a second time and marinate the rabbit heads in the broth till they are served.

Sichuan-Style Heartleaf and Broad Bean Salad
侧耳根拌蚕豆

酸辣鲜香,开胃爽口
sour, spicy, aromatic and appetizing taste

[原 料]
侧耳根(蕺菜)75克 蚕豆150克

[调 料]
蒜蓉3克 葱花10克 花椒粉2克 食盐3克 酱油5克 醋5克 白糖1克 味精3克 红油辣椒30克

[制 作]
1. 侧耳根去除老根、黄叶,洗净后晾干水分。
2. 鲜蚕豆洗净,放入沸水中煮5分钟,捞出凉冷。
3. 食用时将侧耳根、蚕豆和所有调料拌匀即成。

Ingredients
75g heartleaf (Houttuynia cordata), 150g broad beans

Seasonings
3g garlic (finely chopped), 10g chopped scallion, 2g ground roasted Sichuan pepper, 3g salt, 5g soy sauce, 5g vinegar, 1g sugar, 3g MSG, 30g oil-infused chili flakes

Preparation
1. Remove the coarse roots and withered leaves of the heartleaf, rinse and drain.
2. Wash the broad beans and boil in water for 5 minutes. Remove, drain and cool.
3. Mix the heartleaf, broad beans and the seasonings. Blend well and serve.

灯影苕片
Translucent Sweet Potato Chips

色泽金红，酥脆爽口，麻辣回甜，片薄透明
bright brown color; crispy and crunchy chips; spicy and subtly sweet taste; thin and translucent slices

灯影苕片
Translucent Sweet Potato Chips

[原　料]

红苕1个（约200克）　食用油1500克（约耗50克）

[调　料]

食盐2克　白糖2克　辣椒油25克　花椒油1克　味精0.5克　芝麻油5克

[制　作]

1. 红苕洗净去皮，切成长约6厘米、宽约4厘米的长方块，再片成薄片，放入清水中浸泡片刻，捞出沥干水分。
2. 锅置中火上，加入食用油烧至140℃，将红苕片分次放入炸至色红、酥脆时捞出，沥干油分。
3. 辣椒油、花椒油、食盐、白糖、芝麻油、味精放入碗内调成味汁，与苕片拌匀后装盘即成。

Ingredients

1 sweet potato (about 200g), 1500g cooking oil for deep-frying

Seasonings

2g salt, 2g sugar, 25g chili oil, 1g Sichuan pepper oil, 0.5g MSG, 5g sesame oil

Preparation

1. Wash and peel the sweet potato. Cut the potato first into rectangular chunks about 6cm long and 4cm wide, then into thin slices. Soak the slices in water for a while. Remove, rinse and drain.
2. Heat oil in a wok to 140℃ and deep-fry the potato slices till they become brown and crispy. Remove and drain.
3. Mix the chili oil, Sichuan pepper oil, salt, sugar, sesame oil and MSG in a bowl. Add the potato slices, blend well and transfer to a plate.

口口脆

Crunchy Auparagus Lettuce

色泽碧绿，口感爽脆
verdant color; crisp, refreshing and appetizing taste

[原　料]
青笋1000克

[调　料]
食盐5克　柠檬汁50克　白糖20克　芝麻油5克　花椒油5克

[制　作]
青笋切成薄片，加食盐腌渍1小时，用纱布包好压干水分，入冰箱冰镇后取出；加柠檬汁、白糖、芝麻油、花椒油拌匀装盘即成。

Ingredients
1000g asparagus lettuce

Seasonings
5g salt, 50g lemon juice, 20g sugar, 5g sesame oil, 5g Sichuan pepper oil

Preparation
Cut the asparagus lettuce into thin slices, mix with salt and marinate for about 1 hour. Wrap in a piece of cheese cloth, press to dry and then freeze in a freezer. Get the asparagus lettuce out of the freezer, add lemon juice, sugar, sesame oil, Sichuan pepper oil, blend well and transfer to a serving dish.

Hot-and-Sour Fern Root Noodles
酸辣蕨粉

蕨粉滑爽，咸鲜酸辣，鲜香可口
smooth and appetizing noodles; sour and pungent taste

[原　料]
蕨根粉150克

[调　料]
小米辣椒粒20克　野山椒水20克　葱花4克　香菜碎2克　食盐3克　味精2克　生抽10克　醋10克　冷鲜汤20克　芝麻油5克

[制　作]
1. 汤锅置火上，入水烧沸，入蕨根粉煮约3分钟至软时捞出，用清水漂冷。
2. 将小米椒粒、野山椒水、食盐、味精、生抽、醋、冷鲜汤、芝麻油调匀成味汁。
3. 捞出凉透的蕨根粉，沥干水分后放入盘中，浇上调味汁，撒上葱花、香菜碎即成。

Ingredients
150g dried fern root noodles

Seasonings
20g birdseye peppers (finely chopped), 20g brine of pickled tabasco peppers, 4g scallion (finely chopped), 2g coriander (finely chopped), 3g salt, 2g MSG, 10g soy sauce, 10g vinegar, 20g everyday stock, 5g sesame oil

Preparation
1. Bring water in a wok to a boil, blanch the fern root noodles in boiling water for about 3 minutes, remove and rinse in water till cool.
3. Mix birdseye peppers, brine of pickled tabasco peppers, salt, MSG, soy sauce, vinegar, stock and sesame oil. Blend well to make seasoning sauce.
3. Remove the fern root noodles from water, drain and transfer to a plate. Pour the seasoning sauce over the noodles and sprinkle with scallion and coriander.

质地酥脆，爽口化渣，咸甜麻辣香几味兼宜
crispy and crunchy peanuts; a medley of flavors

怪味花仁
Multi-Flavored Peanuts

[原　料]
盐酥花生150克

[调　料]
白糖75克　甜面酱15克　辣椒粉5克　花椒粉2克　食盐1克

[制　作]
1. 盐酥花生搓掉皮衣。
2. 锅内加清水、白糖，用小火熬成浓糖浆，加甜面酱拌匀，再加入食盐、辣椒粉、花椒粉和匀，熄火冷一下；倒入花仁均匀粘裹上糖液，待其翻砂后起锅，晾冷装盘即成。

Ingredients
150g salty crispy peanuts (roasted or fried)

Seasonings
75g sugar, 15g fermented flour paste, 5g ground chilies, 2g ground roasted Sichuan pepper, 1g salt

Preparation
1. Rub off and discard the peanut skins.
2. Heat water in a wok, add sugar and simmer to make thick syrup. Add the fermented flour paste and mix well. Add the ground chilies and ground roasted Sichuan pepper, and blend well. Switch off the flame and allow to cool for a few minutes before adding the peanuts. Stir well to ensure that the peanuts are evenly coated with the syrup. Wait till cool and transfer to a serving dish.

色彩翠绿，酸香脆嫩，姜味浓郁

agreeable verdant color; crisp and tender asparagus beans; refreshingly sour taste with a strong ginger flavor

姜汁豇豆
Asparagus Beans in Ginger Sauce

[原 料]

嫩豇豆200克

[调 料]

姜末8克　食盐2克　味精2克　醋8克　冷鲜汤20克
芝麻油8克

[制 作]

1. 豇豆洗净，入沸水锅中略煮至断生时捞出，放入冷开水中漂冷待用。
2. 将调料调匀成姜汁味汁。
3. 捞出豇豆，沥干水分，切成约5厘米长的段，整齐地装入盘内，淋上姜汁味汁即成。

Ingredients

200g tender asparagus beans

Seasonings

8g ginger (finely-chopped), 2g salt, 2g MSG, 8g vinegar, 20g everyday stock, 8g sesame oil

Preparation

1. Rinse the asparagus beans, boil them in water till al dente, remove and then soak in cold water to cool.
2. Mix the Seasonings to make ginger-flavored sauce.
3. Cut the asparagus beans into 5cm lengths, lay them neatly on a serving dish and then pour over the sauce.

泡椒双耳
Black and White Chili-Pickle-Flavored Funguses

色彩美观,质地脆嫩,泡椒味浓
pleasant and beautiful color; crunchy and tender fungi with pickled chili flavor

[原 料]

水发黑木耳100克 水发白木耳100克

[调 料]

泡菜盐水500克 野山椒末50克 红尖椒粒30克 食盐2克 白醋10克 白糖2克

[制 作]

1. 将各种调料放入大碗调匀成泡椒盐水。
2. 两种木耳用清水浸泡1小时,去蒂后洗净,用清水煮熟,捞出晾凉后放入泡椒盐水中浸泡2小时,捞出后沥水装盘即成。

Ingredients

100g water-soaked Jew's ear funguses, 100g water-soaked snow funguses

Seasonings

500g pickle brine, 50g tabasco peppers (minced), 30g red chili peppers (finely chopped), 2g salt, 10g vinegar, 2g sugar

Preparation

1. Mix the Seasonings in a big bowl, and blend well to make pickled-chili-flavored brine.
2. Soak the two kinds of funguses in water for about one hour, remove any knobbly bits, rinse in water and boil till cooked through. Remove, cool and marinate in the brine for 2 hours. Remove from the brine, drain and transfer to a serving dish.

质脆嫩爽口,清爽不腻,回味甘甜
crispy, crunchy nuts; delicate, aromatic and slightly sweet taste

椒麻桃仁
Jiaoma-Flavor Walnuts

[原　料]
核桃仁200克

[调　料]
椒麻糊30克　食盐2克　味精1克　酱油10克
冷鸡汤40克　芝麻油8克

[制　作]
1. 核桃仁用沸水烫泡约5分钟后捞出,撕去外皮后洗净。
2. 调味碗中加入椒麻糊、食盐、味精、酱油、冷鸡汤、芝麻油调成椒麻味料,再放入核桃仁拌匀,装盘成菜。

Ingredients
200g walnut kernels

Seasonings
30g Jiaoma paste, 2g salt, 1g MSG, 10g soy sauce, 40g cold chicken stock, 8g sesame oil

Preparation
1. Soak the walnut kernels in boiling water for 5 minutes. Scrape off their skins and rinse.
2. Put the Jiaoma paste into a bowl, and add the salt, MSG, soy sauce, chicken stock and sesame oil to make Jiaoma-flavor sauce. Blend in the walnuts, stir well and then transfer to a serving dish.

Buckwheat Noodles with Shredded Chicken
荞面鸡丝

面条滑爽，鲜香微辣

smooth and slithery noodles; tasty and slightly spicy taste

[原 料]

荞麦面条200克 熟鸡肉丝50克

[调 料]

食盐3克 味精2克 胡椒粉2克 酱油20克 辣椒油40克 花椒粉2克 葱花10克 鲜汤100克

[制 作]

1. 碗中加入食盐、花椒粉、胡椒粉、酱油、味精、葱花、辣椒油、鲜汤调成味汁。
2. 锅置旺火上，掺入清水烧沸，下荞麦面条煮熟，捞出后放入味汁碗中，撒上鸡丝即成。

Ingredients

200g buckwheat noodles, 50g shredded precooked chicken

Seasonings

3g salt, 2g MSG, 2g ground white pepper, 20g soy sauce, 40g chili oil, 2g ground roasted Sichuan pepper, 10g scallion (finely chopped), 100g everyday stock

Preparation

1. Mix salt, ground roasted Sichuan pepper, ground white pepper, soy sauce, MSG, scallion, chili oil and stock in a serving bowl to make seasoning sauce.
2. Heat water in a wok over a high flame, bring to a boil and add the noodles. Boil till the noodles are cooked through. Transfer the noodles to the bowl. Sprinkle with shredded chicken.

清香脆嫩，芝麻酱香
fresh, delicate and crisp asparagus lettuce;
sesame paste aroma

麻酱凤尾

Asparagus Lettuce with Sesame Paste

[原　料]

青笋尖200克

[调　料]

芝麻酱30克　芝麻油15克　食盐1克　味精2克
白糖5克　酱油10克

[制　作]

1. 青笋尖两端修切整齐成长约13厘米的段，洗净后晾干水分，把嫩茎一端削成青果尖形并切开成四瓣，整齐地装入餐盘内。
2. 芝麻酱、食盐、白糖、酱油、味精、芝麻油调匀成味汁，装入调味碟内，与笋尖一起上桌即成。

Ingredients

200g asparagus lettuce tips (the leafy parts)

Seasonings

30g sesame paste, 15g sesame oil, 1g salt, 2g MSG, 5g sugar, 10g soy sauce

Preparation

1. Trim the two ends of the asparagus lettuce tips so that they are about 13cm long, rinse and drain. Cut the stem end first into the shape of an olive and then divide each lettucetip into four parts. Lay neatly on a serving dish.
2. Mix the Seasonings to make dipping sauce, and transfer to a saucer. Serve the dipping sauce with the asparagus lettuce.

豌豆酥软，咸甜酸辣兼备，姜葱蒜味浓郁
crisp crust and soft interior; a mixture of salty, sweet, sour and slightly hot tastes; appealing aroma of ginger, garlic and scallion

鱼香豌豆
Peas in Fish-Flavor Sauce

[原　料]

青豌豆300克　食用油1000克（约耗50克）

[调　料]

泡椒末30克　姜末5克　蒜末10克　葱花30克　辣椒油25克　芝麻油25克　酱油20克　醋20克　白糖15克　食盐3克　味精2克

[制　作]

1．将调料混合后调成鱼香味汁。
2．锅置旺火上，入食用油加热至160℃，下青豌豆炸至浮起、酥脆，去掉豌豆壳，捞出豌豆，沥干炸油，趁热淋上鱼香味汁拌匀即成。

Ingredients

300g peas, 1000g cooking oil for deep-frying

Seasonings

30g pickled chilies (finely chopped), 5g ginger (finely chopped), 10g garlic (finely chopped), 30g scallion (finely chopped), 25g chili oil, 25g sesame oil, 20g soy sauce, 20g vinegar, 15g sugar, 3g salt, 2g MSG

Preparation

1. Mix the Seasonings to make the fish-flavor sauce.
2. Heat oil in a wok over a high flame to 160℃, toss in the peas to deep-fry till they float on the oil surface and become crisp. Remove and discard the floating pea crusts. Remove the peas from the oil with a perforated spoon, and drain. Pour the fish-flavor sauce over the peas while thay are still hot, and then blend well.

鱼香豌豆
Peas in Fish-Flavor Sauce

Sichuan (China) Cuisine in Both Chinese and English

川菜
(中英文标准对照版)

热 菜
Hot Dishes

第二篇

汁红芡亮，咸鲜醇厚，鱼质柔嫩
brown and lustrous sauce; rich flavors; tender
and soft abalone

红烧鲍鱼
Red-Braised Abalone

[原 料]

水发鲍鱼500克 水发香菇50克 冬笋50克 熟火腿50克 菜心100克 化猪油25克

[调 料]

食盐3克 姜片10克 葱段15克 料酒20克 味精3克 胡椒粉1克 酱油15克 水淀粉15克 鲜汤700克 化鸡油20克

[制 作]

1. 鲍鱼片成片（或剞花刀），放入加有料酒、味精的沸鲜汤中喂味；水发香菇、冬笋、熟火腿切成片；菜心洗净。
2. 锅中放化猪油烧至120℃时，放入姜片、葱段炒香，掺入鲜汤烧沸，捞出姜、葱，放入食盐、味精、料酒、胡椒粉、酱油、水发香菇、冬笋、火腿、菜心烧熟至入味，捞出后装盘垫底；将鲍鱼入锅烧至入味，捞出放在辅料上；锅中汤汁用水淀粉勾成清二流芡，起锅前淋上化鸡油推匀，浇淋在鲍鱼上即成。

Ingredients

500g water-soaked abalone, 50g water-soaked shiitake mushrooms, 50g winter bamboo shoots, 50g pre-cooked ham, 100g tender leafy vegetables, 25g lard

Seasonings

3g salt, 10g ginger (sliced), 15g scallion (cut into sections), 20g Shaoxing cooking wine, 3g MSG, 1g ground white pepper, 15g soy sauce, 15g cornstarch-water mixture, 700g everyday stock, 20g chicken oil

Preparation

1. Cut the abalone into slices, and transfer into a potful of boiling stock with the MSG and Shaoxing cooking wine so that the abalone absorbs the flavor of the stock. Cut the shiitake mushrooms, winter bamboo shoots and ham into slices.
2. Heat the lard in a wok to 120℃, add the ginger and scallion, and stir-fry to bring out the fragrance. Add the stock, bring to a boil, remove the ginger and scallion, and then add salt, MSG, Shaoxing cooking wine, ground white pepper, soy sauce, shiitake mushrooms, bamboo shoots, ham and tender leafy vegetables. Bring to a boil, simmer till fully cooked, and then remove the solid ingredients with a slotted spoon and lay them in a serving dish. Add the abalone to the remaining liquid, bring to a boil and simmer to allow it to absorb the flavors. Lay the abalone over the other ingredients in the serving dish. Add the cornstarch-water mixture and the chicken oil to the wok, stir well and pour over the abalone.

红烧鲍鱼 Red Braised Abalone

宫保龙虾球
Lobster Balls

色泽棕红，虾肉细嫩，腰果酥脆，咸鲜甜酸，麻辣味浓
bright brown color; tender lobster; crunchy cashew nuts; a complication of sour, sweet, pungent, tingling and aromatic tastes

宫保龙虾球
Gongbao Lobster Balls

[原　料]
龙虾1只（约750克）　酥腰果60克　食用油600克（约耗50克）

[调料A]
干辣椒段10克　花椒2克　姜片8克　蒜片8克　葱丁20克

[调料B]
食盐1克　料酒5克　蛋清10克　干淀粉10克

[调料C]
食盐1克　料酒5克　酱油7克　醋10克　白糖10克　味精1克　水淀粉15克　鲜汤30克

[制　作]
1. 将调料C调匀成芡汁。
2. 龙虾经初加工后取肉，切成约2厘米大小的丁，与调料B拌匀，入80℃食用油中滑熟捞出。
3. 锅内留油烧至120℃，下调料A炒香，再下龙虾肉炒散籽，烹入芡汁，放入腰果炒匀，起锅装盘即成。

Ingredients
1 lobster (about 750g), 60g crispy cashew nuts, 600g cooking oil for deep-frying

Seasonings A
10g dried chilies (cut into sections), 2g Sichuan pepper, 8g ginger (sliced), 8g garlic (sliced), 20g scallion (chopped)

Seasonings B
1g salt, 5g Shaoxing cooking wine, 10g egg white, 10g cornstarch

Seasonings C
1g salt, 5g Shaoxing cooking wine, 7g soy sauce, 10g vinegar, 10g sugar, 1g MSG, 15g cornstarch-water mixture, 30g everyday stock

Preparation
1. Mix Seasonings C to make thickening sauce.
2. Kill and rinse the lobster. Cut the lobster meat into 2cm³ cubes. Mix with Seasonings B and blend well. Slide in 80℃ oil to fry till cooked through, and remove.
3. Leave a little oil in the wok and heat to 120℃. Add Seasonings A, stir-fry till aromatic, and then add the lobster meat, stirring so that the lobster cubes do not stick together. Pour in the thickening sauce, add the cashew nuts and blend well. Remove from the stove and transfer to a serving dish.

色泽红亮，咸甜酸辣兼备，葱姜蒜香浓郁
bright color; a combination of salty, sweet, sour and slightly hot tastes; scrumptious aroma of scallion, ginger and garlic

鱼香龙虾
Lobster in Fish-Flavor Sauce

[原 料]
龙虾1只（约重1500克） 食用油750克（约耗70克）

[调 料]
泡辣椒末40克 姜米10克 蒜米15克 葱花20克 食盐3克 料酒10克 白糖25克 醋20克 酱油7克 味精2克 鲜汤50克 蛋清淀粉70克 水淀粉10克

[制 作]
1. 龙虾经初加工后取肉，切成约2厘米见方的丁；将食盐、料酒、白糖、醋、酱油、味精、鲜汤、水淀粉兑成芡汁。
2. 龙虾肉加入食盐、料酒、蛋清淀粉拌匀，入100℃的油中滑散，刚熟时捞出。
3. 锅中留油，放入泡辣椒末、姜米、蒜米、葱花炒香，入龙虾肉炒匀，倒入芡汁，待收汁亮油后装盘即成。

Ingredients
1 lobster (1500g), 750 cooking oil for deep-frying

Seasonings
40g pickled chilies (finely chopped), 10g ginger (finely chopped), 15g garlic (finely chopped), 20g scallion (finely chopped), 3g salt, 10g Shaoxing cooking wine, 25g sugar, 20g vinegar, 7g soy sauce, 2g MSG, 50g everyday stock, 70g mixture of egg white and cornstarch, 10g cornstarch-water mixture

Preparataion
1. Kill and clean the lobster. Cut lobster meat into 2cm³ cubes. Mix salt, Shaoxing cooking wine, sugar, vinegar, soy sauce, MSG, stock and cornstarch-water mixture to make thickening sauce.
2. Mix the lobster cubes with salt, Shaoxing cooking wine and the mixture of egg white and cornstarch, blend well and deep-fry in 100℃ oil, stirring constantly to make sure that the lobster cubes separate. Remove when they are just cooked.
3. Heat some oil in the wok, add pickled chilies, ginger, garlic and scallion, and stir-fry to bring out the aroma. Add lobster cubes, stir well and pour in the thickening sauce. Wait till the sauce becomes thick and lustrous, and transfer to a serving dish.

干烧大虾
Dry-Braised Prawns

色泽红亮，质地脆嫩，咸鲜微辣，口感回甜
bright colors; tender prawn meat; salty, savoury, slightly hot and sweet taste

干烧大虾
Dry-Braised Prawns

[原 料]
大虾400克 猪肉末100克 食用油1000克（约耗70克）

[调 料]
泡辣椒段8根 姜米10克 蒜米10克 葱段50克 碎米芽菜20克 食盐3克 味精1克 料酒15克 酱油5克 白糖5克 芝麻油5克

[制 作]
1. 大虾去尽虾须、沙线后洗净，入150℃的食用油中炸至酥香、色红时捞出。
2. 锅置旺火上，入食用油烧至120℃，下肉末炒香，再下酱油、食盐、料酒炒酥，续入泡辣椒段、葱段、姜米、蒜米炒香后掺汤，再放入大虾、碎米芽菜、食盐、料酒、白糖，烧至入味、汁干、亮油后下味精、芝麻油，最后捡去部分葱段和泡辣椒后装盘即成。

Ingredients
400g prawns, 100g pork mince, 1000g cooking oil for deep-frying

Seasonings
8 lengths of pickled chilies, 10g ginger (finely chopped), 10g garlic (finely chopped), 50g scallion (cut into sections), 20g yacai (preserved mustard stems minced), 3g salt, 1g MSG, 15g Shaoxing cooking wine, 5g soy sauce, 5g sugar, 5g sesame oil

Preparation
1. De-palp and devein the prawns. Wash them thoroughly and fry in 150℃ oil till red and aromatic.
2. Heat oil in a wok over a high flame to 120℃, add the pork mince and stir-fry till aromatic. Add the soy sauce, salt and Shaoxing cooking wine, and continue to stir-fry till the mince becomes crisp. Stir in the pickled chilies, scallion, ginger and garlic to bring out the aroma. Add the stock, prawns, minced yacai, salt, Shaoxing cooking wine and sugar. Bring to a boil and then simmer till the soup has been fully absorbed by the prawns and the oil is clear. Season with MSG and sesame oil. Remove and discard some of the scallion and chili pickles, and transfer to a serving dish.

Shrimps with Jade-Colored Broad Beans

翡翠虾仁

白绿相间，虾仁质嫩，蚕豆质软，咸鲜清香
contrasting colors of white and green; tender shrimps and beans; salty and delicate taste

[原 料]

虾仁200克 蚕豆100克 食用油1000克（约耗70克）

[调 料]

食盐2克 胡椒粉1克 味精1克 料酒10克 鲜汤20克 蛋清淀粉70克 水淀粉15克

[制 作]

1. 蚕豆氽水断生；虾仁与料酒、食盐、蛋清淀粉拌匀。
2. 食盐、胡椒粉、味精、鲜汤、水淀粉调匀成咸鲜调味芡汁。
3. 锅中放油烧至100℃，放虾仁滑熟，滗去余油，入蚕豆略炒，倒入调味芡汁，收汁亮油后起锅装盘即成。

Ingredients

200g shelled shrimps, 100g broad beans, 1000g cooking oil for deep-frying

Seasonings

2g salt, 1g ground white pepper, 1g MSG, 10g Shaoxing cooking wine, 20g everyday stock, 70g mixture of egg white and cornstarch, 15g cornstarch-water mixture

Preparation

1. Boil the broadbeans in water till just cooked. Combine the shrimps, Shaoxing cooking wine, salt and the mixture of egg white and cornstarch, and mix well.
2. Mix salt, ground white pepper, MSG, stock and cornstarch-water mixture to make the thickening sauce.
3. Heat the oil in a wok to 100℃ and deep-fry the shrimps till cooked through. Drain off the excess oil, add the broad beans, and stir-fry briefly in the remaining oil. Pour in the thickening sauce, and wait till the liquid becomes thick and lustrous. Transfer to a serving dish.

典故

翡翠本指硬玉，多为绿色、蓝绿色或白色中带绿色斑纹。本菜中的蚕豆颜色翠绿，与翡翠颜色相似，故美其名曰"翡翠虾仁"。

Note

Jade is a precious stone with a delicate green color. The broadbeans in this dish are similar in color to jade, which is why the dish is so named.

麻辣鲜香，味浓汁爽
spicy, pungent, numbing and savoury taste

盆盆虾

Penpen Prawns (Spicy Prawns in a Basin)

[原 料]
鲜活大虾600克 豆芽300克 食用油1500克（约耗200克）

[调 料]
郫县豆瓣50克 姜片20克 葱段2克 干辣椒节50克 花椒20克 辣椒粉20克 胡椒粉2克 料酒20克 白糖5克 食盐3克 味精4克 芝麻油10克 老油100克 熟芝麻10克 葱花10克 鲜汤300克

[制 作]
1. 豆芽入沸水锅氽熟，捞入盆中垫底。锅中放食用油烧至160℃，放入鲜虾炸1分钟捞出。
2. 锅中放食用油烧至120℃，放入郫县豆瓣、干辣椒节、花椒、姜片、葱段、辣椒粉炒香，下鲜虾、料酒、胡椒粉、白糖、食盐、老油炒匀，再加入鲜汤、味精、芝麻油烧至虾肉入味，连味汁一起倒入盛豆芽的盆中，撒上葱花和熟芝麻即成。

Ingredients
600g live prawns, 300g bean sprouts, 1500g cooking oil for deep-frying

Seasonings
50g Pixian chili bean paste, 20g ginger, 2g scallion, 50g dried chilies (cut into sections), 20g Sichuan pepper, 20g ground chilies, 2g ground white pepper, 20g Shaoxing cooking wine, 5g sugar, 3g salt, 4g MSG, 10g sesame oil, 100g hot pot oil, 10g roasted sesame seeds, 10g scallion (finely chopped), 300g everyday stock

Preparation
1. Blanch the bean sprouts in water and transfer to a serving basin. Heat the oil in a wok to 160℃, deep-fry the prawns for one minute, and then remove.
2. Heat some oil in a wok to 120℃, add Pixian chili bean paste, dried chilies, Sichuan pepper, scallion and ground chilies, and stir-fry till aromatic. Add the prawns, Shaoxing cooking wine, ground white pepper, sugar, salt and hot pot oil, stir well and add the stock, MSG and sesame oil. Braise till the prawns have absorbed the flavors of the seasonings. Pour the contents of the wok into the serving basin. Sprinkle with scallion and sesame seeds.

典故
此菜因盛器为陶瓷盆或不锈钢盆而得名，虾吃完还可以把汤汁重新倒回锅内，下豆腐、粉皮、魔芋等其他原料再煲一锅美味。

Note
The dish is so named because the container of the dish is a steel or earthen basin, which is called Penpen in Sichuan dialect. After eating the prawns, return the sauce into a wok, and add tofu, steamed bean jelly and konjak curd. Boil till they are fully cooked, and serve.

干烧辽参

Dry-Braised Liaoning Sea Cucumber

色泽棕红，辽参软糯，咸鲜味浓
bright brown color; soft, tender and glutinous Liaoning sea cucumber; salty, aromatic and savory taste.

干烧辽参
Dry-Braised Liaoning Sea Cucumber

[原料]
辽参150克 猪肉末30克 食用油50克

[调料]
碎米芽菜10克 姜米5克 葱段25克 蒜米5克 味精2克 食盐3克 酱油5克 芝麻油5克 鲜汤1000克

[制作]
1. 辽参涨发后入锅内，加入鲜汤，用小火煨至柔软。
2. 锅置旺火上，入食用油烧至120℃，放猪肉末和碎米芽菜炒香，续下姜米、葱段、蒜米、食盐、酱油炒香，再加鲜汤和辽参烧至汁干、亮油，最后下味精、芝麻油，出锅装盘即成。

Ingredients
150g Liaoning sea cucumber, 30g pork mince, 50g cooking oil

Seasonings
10g minced yacai (preserved mustard stems), 5g ginger (finely chopped), 25g scallion (finely chopped), 5g garlic (finely chopped), 2g MSG, 3g salt, 5g soy sauce, 5g sesame oil, 1000g everyday stock

Preparation
1. Soak the sea cucumber in water to reconstitute. Transfer to a wok, add the stock and simmer till soft.
2. Heat oil in a wok over a high flame to 120℃, add the pork mince and yacai, and stir-fry to bring out the fragrance. Add the ginger, garlic, scallions, salt and soy sauce, and stir-fry till aromatic. Add the stock and the sea cucumber, and braise till the liquid is much reduced and the oil clear. Add MSG and sesame oil and transfer to a serving plate.

Home-Style Sea Cucumber
家常海参

色泽红亮，质地软糯，咸鲜微辣，香浓醇厚

bright and lustrous color; soft and slithery meat; salty, savoury and slightly hot taste; refreshing and aromatic smell

[原 料]

水发海参300克 猪肉臊50克 黄豆芽75克 蒜苗花15克 食用油75克

[调 料]

姜片10克 葱段15克 郫县豆瓣40克 食盐2克 味精2克 酱油5克 料酒20克 芝麻油10克 水淀粉18克 鲜汤150克

[制 作]

1. 将水发海参片成长10厘米、上厚下薄的斧楞片。
2. 锅置火上，入少量食用油烧至120℃，入姜片、葱段炒香，再加鲜汤烧沸，倒入盛海参的碗内喂味。
3. 锅置火上，入食用油烧至150℃，下黄豆芽、食盐炒至断生，起锅装入盘内垫底。
4. 锅置火上，入食用油烧至120℃，下郫县豆瓣炒香，续加鲜汤、海参、猪肉臊、食盐、料酒、酱油，用中火烧至入味，再加味精、蒜苗花，用水淀粉勾芡，淋入芝麻油，起锅浇盖在黄豆芽上即成。

Ingredients

300g water-soaked sea cucumber, 50g stir-fried pork mince, 75g soy bean sprouts, 15g baby leeks (finely chopped), 75g cooking oil

Seasonings

10g ginger (sliced), 15g scallion (cut into sections), 40g Pixian chili bean paste, 2g salt, 2g MSG, 5g soy sauce, 20g Shaoxing cooking wine, 10g sesame oil, 18g cornstarch-water mixture, 150g everyday stock

Preparation

1. Cut the water-soaked sea cucumber into 10cm-long slices that are thick on one end and thin on the other.
2. Heat some oil in a wok to 120°C, add the ginger and scallion, and stir-fry to bring out the fragrance. Add some stock and bring to a boil. Pour the boiling stock into the bowl containing the sea cucumber to marinate.
3. Heat some oil in the wok to 150°C, add the soy bean sprouts and salt, and sauté till al dente. Remove and transfer to a serving plate.
4. Heat some oil in the wok to 120°C, add the Pixian chili bean paste, and stir-fry till the oil becomes reddish and aromatic. Add some stock, the sea cucumber, stir-fried pork mince, salt, Shaoxing cooking wine and soy sauce. Braise over a medium flame so that the sea cucumber pieces absorb the flavors. Add the MSG and leeks, thicken with the cornstarch-water mixture, and drizzle with the sesame oil. Remove from fire and pour over the soy bean sprouts.

汤清呈浅棕色，味清鲜酸辣，质地柔软爽口

appealing brownish color; soft and slithery sea cucumber; sour and pungent taste

酸辣海参

Hot-and-Sour Sea Cucumber

[原　料]

水发海参500克　冬笋100克　鸡蛋1个　清汤750克

[调　料]

食盐4克　胡椒粉4克　味精2克　料酒10克　醋20克　葱花5克　姜末15克　鲜汤1000克

[制　作]

1. 将水发海参片成2毫米厚的薄片；冬笋切成2厘米见方的薄片；鸡蛋煮熟，去蛋黄，切成2厘米见方的薄片。
2. 锅置火上，掺鲜汤，放食盐、料酒烧沸，入海参氽煮2次；冬笋片入沸水锅中氽水后捞出。
3. 取大汤碗1个，放入醋、味精、胡椒粉、葱花、鸡蛋片、冬笋片、海参片。
4. 锅置火上，掺清汤，放入食盐、姜末烧沸，舀入汤碗内即成。

Ingredients

500g water-soaked sea cucumber, 100g winter bamboo shoots, 1 egg, 750g consomme

Seasonings

4g salt, 4g ground white pepper, 2g MSG, 10g Shaoxing cooking wine, 20g vinegar, 5g scallion (finely chopped), 15g ginger (finely chopped), 1000g everyday stock

Preparation

1. Cut the sea cucumber into 2mm-thick slices. Cut the bamboo shoots into (2cm)² slices. Boil the egg, remove the yolk and then cut the white into (2cm)² slices.
2. Heat an oven over a flame, add stock, salt and Shaoxing cooking wine, bring to a boil and blanch the sea cucumber twice in the stock. Blanch the bamboo shoot slices in water.
3. Mix salt, vinegar, MSG, ground white pepper, scallion, egg slices, bamboo shoot slices and sea cucumber slices in a large soup bowl.
4. Put an oven over a flame, add consomme, salt and ginger, bring to a boil and pour into the soup bowl.

白汁鱼肚卷

汁色乳白，菜色浅黄，鲜嫩可口

Fish Maw Rolls in Milky Sauce

creamy sauce; delicate and appetizing taste

[原　料]

油发鱼肚500克　鱼糁150克　菜心200克　化猪油30克

[调　料]

姜片10克　葱段15克　食盐2克　味精1克　胡椒粉1克　料酒10克　化鸡油10克　水淀粉20克　蛋清淀粉50克　奶汤250克

[制　作]

1. 鱼肚去油后切成长5厘米、宽2.5厘米、厚0.5厘米的片，用鲜汤喂味后挤干水分。
2. 将鱼肚平铺在菜墩上，先抹上蛋清淀粉，再抹上一层鱼糁，裹成卷筒，入笼蒸熟后取出，晾冷后从中间切断，定蒸碗，入笼蒸热，翻扣在炒熟的菜心上。
3. 锅中放化猪油烧至120℃，放姜片、葱段炒香，掺入奶汤，加食盐、胡椒粉、料酒、味精烧沸，用水淀粉勾成清二流芡，浇淋在鱼肚卷上即成。

Ingredients

500g oil-soaked fish maw, 150g fish mince, 200g tender leafy vegetables, 750g stock, 30g lard

Seasonings

10g ginger (sliced), 15g scallion (cut into sections), 2g salt, 1g MSG, 1g ground white pepper, 10g Shaoxing cooking wine, 10g chicken oil, 20g cornstarch-water mixture, 50g mixture of egg white and cornstarch, 250g milky stock

Preparation

1. Remove the oil from the fish maw and cut into slices about 5cm long, 2.5cm wide and 0.5cm thick. Marinate the fish maw slices in some everyday stock, and then drain.
2. Flatten the fish maw slices on a chopping board, smear them first with the mixture of egg white and cornstarch and then with the fish mince, roll up, and steam in a steamer till cooked through. Remove the fish maw rolls and cool. Halve the rolls, transfer to a steaming bowl, steam in a steamer till hot, and then turn the bowl upside down to pour the rolls onto a bed of stir-fried vegetables.
3. Heat the lard in a wok to 120℃, add the ginger and scallion, and stir-fry till aromatic. Add the milky stock, salt, ground white pepper, Shaoxing cooking wine and MSG, bring to a boil, add the cornstarch-water mixture to thicken the sauce, and pour over the rolls.

Turbot in Pepper-Flavored Sauce
椒汁多宝鱼

鱼肉细嫩，咸鲜带辣，有青尖椒独具的清香

tender fish; salty, refreshing and spicy taste; the fragrance of green chili peppers

[原 料]

多宝鱼1尾（约750克） 青尖椒200克 红尖椒10克 食用油30克

[调 料]

姜片10克 葱段15克 食盐4克 料酒15克 味精2克 酱油10克 芝麻油5克 鲜汤50克

[制 作]

1. 青尖椒、红尖椒切成小丁。
2. 多宝鱼初加工后用姜片、葱段、食盐、料酒码味15分钟，入笼用旺火蒸8分钟后拣出姜片、葱段。
3. 锅中放食用油烧至140℃，放青尖椒、红尖椒丁炒熟，入食盐、味精、酱油、芝麻油、鲜汤调好味，起锅浇淋在鱼上即成。

Ingredients

1 turbot (about 750g), 200g green chili peppers, 10g red chili peppers, 30g cooking oil

Seasonings

10g ginger (sliced), 15g scallion (cut into sections), 4g salt, 15g Shaoxing cooking wine, 2g MSG, 10g soy sauce, 5g sesame oil, 50g everyday stock

Preparation

1. Chop the green and red chili peppers first into thin strips, and then into tiny squares.
2. Kill and clean the turbot, and marinate in the mixture of ginger, scallion, salt and Shaoxing cooking wine for 15 minutes. Transfer to a steamer and steam over a high flame for 8 minutes. Remove and discard the ginger and scallion.
3. Heat oil in a wok to 140℃, add the chopped peppers, and stir-fry till cooked through. Add salt, MSG, soy sauce, sesame oil and stock to make the seasoning sauce. Pour the sauce over the fish.

成菜美观，细嫩柔软，汤浓味鲜
beautifully arranged dumplings and fish maw;
tender and soft taste; creamy and savoury sauce

菠饺鱼肚
Spinach-Flavored Dumplings with Fish Maw

[原　料]

水发鱼肚400克　熟鸡片50克　熟火腿片50克　菜心50克　面粉200克　菠菜汁20克　猪肉馅200克　化猪油30克

[调　料]

食盐3克　胡椒粉1克　料酒15克　味精2克　姜片10克　葱段15克　化鸡油5克　奶汤300克　鲜汤300克

[制　作]

1．水发鱼肚切成4厘米大的薄片，用鲜汤喂味。
2．将面粉、菠菜汁和适量清水和匀调成面团，再包上肉馅，做成24个饺子，入沸水煮熟。
3．锅中放化猪油烧至120℃，入姜片、葱段炒香，掺入奶汤，加食盐、胡椒粉、料酒、味精、化鸡油烧沸，再入熟鸡片、熟火腿片、菜心煮熟，捞出放在盘中垫底；锅中加鱼肚烧入味，捞出盖在辅料上，四周放上煮熟的饺子，浇上锅中的奶汤汁即成。

Ingredients

400g water-soaked fish maw, 50g precooked chicken (sliced), 50g precooked ham (sliced), 50g tender leafy vegetables, 200g plain flour, 20g spinach juice, 200g pork stuffing, 30g lard

Seasonings

3g salt, 1g ground white pepper, 15g Shaoxing cooking wine, 2g MSG, 10g ginger (sliced), 15g scallion (cut into sections), 5g chicken oil, 300g milky stock, 300g everyday stock

Preparation

1. Cut the fish maw into 4cm-long slices and marinate in stock.
2. Mix flour, spinach juice and water to make a dough. Wrap the pork stuffing in dough to make 24 dumplinngs. Boil the dumplings in water till fully cooked.
3. Heat lard in a wok to 120℃, add ginger, scallion, milky stock, salt, ground white pepper, Shaoxing cooking wine, MSG and chicken oil, bring to a boil, and then add chicken slices, ham slices and vegetables. Bring to a boil and simmer till cooked through. Remove the solid ingredients from the wok with a slotted spoon and transfer to a serving dish. Add the fish maw to the wok, braise to let it absorb the flavors and then transfer onto the ingredients on the serving dish. Stack the dumplings around the fish maw and pour the sauce in the wok over the ingredients.

家常鱿鱼
Home-style Squid

色泽红亮，咸鲜微辣，质地软糯

bright and appealing color; salty, moreish and slightly spicy taste; tender and soft squid

[原　料]

水发鱿鱼500克　黄豆芽150克　食用油80克

[调　料]

郫县豆瓣30克　姜米10克　蒜花16克　食盐1克　料酒10克　味精1克　酱油10克　醋5克　白糖5克　水淀粉25克　鲜汤300克

[制　作]

1．水发鱿鱼片成片，用鲜汤喂味后沥干水分备用。
2．黄豆芽去两头后洗净，入热油锅中加食盐炒至断生，装入盘中垫底。
3．锅中放食用油烧至120℃，放入郫县豆瓣、姜米、蒜花炒香，加入鲜汤、食盐、料酒、酱油、醋、白糖烧沸出味，用水淀粉勾成二流芡，放入沥干水分的鱿鱼烧热，连汁盛在豆芽上即成。

Ingredients

500g water-soaked squid, 150g soy bean sprouts, 80g cooking oil

Seasonings

30g Pixian chili bean paste, 1g ginger (finely chopped), 16g garlic (finely chopped), 1g salt, 10g Shaoxing cooking wine, 1g MSG, 10g soy sauce, 5g vinegar, 5g sugar, 25g cornstarch-water mixture, 300g everyday stock

Preparation

1. Cut the squid into slices and marinate in the stock.
2. Remove the two ends of the soy bean sprouts, rinse, and stir-fry in a wok till al dente. Add some salt during the process of stir-frying. Transfer to a serving dish.
3. Heat oil in a wok to 120°C, add Pixian chili bean paste, ginger and garlic, and stir-fry to bring out the aroma. Add the stock, salt, Shaoxing cooking wine, soy sauce, vinegar and sugar, bring to a boil, and add the cornstarch-water mixture and drained squid slices to braise till sizzling hot. Pour the contents in the wok over the soy bean sprouts.

Home-style Fish Snouts
家常鱼唇

色泽红亮，味浓香辣，质地软糯
bright appealing color; aromatic and spicy taste; tender fish snouts

[原 料]
水发鱼唇300克 菜心150克 食用油80克

[调 料]
郫县豆瓣30克 姜米10克 蒜花16克 食盐1克 料酒10克 味精1克 酱油8克 白糖5克 水淀粉25克 鲜汤300克

[制 作]
1. 水发鱼唇片成片，用鲜汤喂味后备用。
2. 菜心入热油锅中加食盐炒至断生，装入盘中垫底。
3. 锅中放食用油烧至120℃，入郫县豆瓣、姜米、蒜花炒香，入鲜汤、食盐、料酒、酱油、白糖、鱼唇烧至入味，用水淀粉勾成二流芡，起锅盛入菜心上即成。

Ingredients
300g water-soaked fish snouts, 150g tender leafy vegetables, 80g cooking oil

Ingredients
30g Pixian chili bean paste, 10g ginger (finely-chopped), 16g garlic (finely-chopped), 1g salt, 10g Shaoxing cooking wine, 1g MSG, 8g soy sauce, 5g sugar, 25g cornstarch-water mixture, 300g everyday stock

Preparation
1. Cut the fish snouts into slices and marinate in the stock.
2. Stir-fry the vegetables till al dente, add some salt and transfer to a serving dish.
3. Heat oil in a wok to 120℃, add the chili bean paste, ginger and garlic, and then stir-fry to bring out the aroma. Add the stock, salt, Shaoxing cooking wine, soy sauce, sugar and fish snouts, and braise so that the fish snouts absorb the flavors. Add cornstarch-water mixture to thiken the sauce and pour over the vegetables.

Dry-Fried Squid Slivers
干煸鱿鱼丝

色泽金黄，干香味美

golden brown; aromatic and savoury taste

[原 料]

干鱿鱼200克 猪肉150克 绿豆芽75克 食用油1000克（约耗70克）

[调 料]

食盐3克 料酒20克 酱油10克 白糖5克 味精1克 芝麻油5克

[制 作]

1. 干鱿鱼用热水泡软，切成二粗丝，放入170℃的食用油中炸成金黄色时捞出；猪肉切成二粗丝。
2. 锅中放食用油烧至160℃，放入肉丝反复煸炒至香酥，入鱿鱼丝、豆芽炒熟，再入食盐、料酒、酱油、白糖、味精、芝麻油炒匀即成。

Ingredients

200g dried squid, 150g pork, 75g mung bean sprouts, 1000g cooking oil for deep-frying

Seasonings

3g salt, 20g Shaoxing cooking wine, 10g soy sauce, 5g sugar, 1g MSG, 5g sesame oil

Preparation

1. Soak the dried squid in hot water till soft, cut into slivers about 0.3cm in diameter and 10cm in length, and deep-fry in 170℃ oil till golden brown. Cut the pork into slivers about 0.3cm in diameter and 10cm in length.
2. Heat oil in a wok to 160℃, stir-fry the pork slivers till aromatic and crispy, add squid slivers and bean sprouts, and stir-fry till fully cooked. Add the salt, Shaoxing cooking wine, soy sauce, sugar, MSG and sesame oil, stir well and transfer to a serving dish.

红白分明，油色红亮，质地脆嫩，咸鲜微辣
contrasting colors of white and red; tender and savoury cuttlefish; salty, delectable and slightly pungent taste

泡椒墨鱼仔

Pickled-Chili-Flavored Tiny Cuttlefish

[原　料]
墨鱼仔500克　泡灯笼椒300克　泡辣椒油100克

[调料A]
泡辣椒末30克　葱白20克　姜米10克　蒜米20克

[调料B]
食盐2克　料酒20克　醋10克　胡椒粉2克　味精1克　鸡精2克　醪糟汁30克　花椒油5克　芝麻油5克　鲜汤100克　水淀粉30克

[制　作]
1．将墨鱼仔加少量醋洗涤干净，入沸水中汆水后备用。
2．锅中放泡辣椒油50克烧至120℃，放入调料A炒香，掺鲜汤烧沸出味，沥去料渣成味汁。
3．锅中放泡辣椒油50克烧至120℃，放入泡灯笼椒炒香，续入制好的味汁，下调料B和墨鱼仔，待汤汁浓稠后起锅装盘而成。

Ingredients
500g tiny cuttlefish, 300g pickled bell peppers, 100g pickled chili oil

Seasonings A
30g pickled chilies (minced), 20g scallion (white part only), 10g ginger (minced), 20g garlic (finely chopped)

Seasonings B
2g salt, 20g Shaoxing cooking wine, 10g vinegar, 2g ground white pepper, 1g MSG, 2g chicken essence granules, 30g fermented glutinous rice wine, 5g Sichuan pepper oil, 5g sesame oil, 100g everyday stock, 30g cornstarch-water mixture

Preparation
1. Add some vinegar to water to help clean the tiny cuttlefish, and then blanch the cuttlefish in boiling water.
2. Bring 50g pickled chili oil in a wok to 120℃, add Seasonings A and stir-fry to bring out the aroma. Pour in the stock, bring to a boil and simmer to bring out the flavors. Remove the scums, and the remaining soup will be used as seasoning sauce.
3. Bring 50g pickled chili oil in a wok to 120℃, add pickled bell peppers, and stir-fry to bring out the aroma. Add the seasoning sauce, Seasonings B and the tiny cuttlefish. Wait till the sauce becomes thick, and transfer to a serving dish.

色白形美，荔枝味浓，柔软嫩脆

exquisite shape and pure color; tender and springy squid; a mixture of salty, slightly sweet and sour taste

荔枝鱿鱼卷
Lichi-Flavor Squid Rolls

[原 料]
鲜鱿鱼300克 食用油50克

[调 料]
姜片6克 蒜片10克 马耳葱15克 食盐2克 白糖20克 醋15克 味精1克 料酒10克 胡椒粉1克 芝麻油5克 水淀粉10克

[制 作]
1. 鲜鱿鱼用刀剞成荔枝花形，入沸水中氽一水。
2. 食盐、白糖、醋、味精、料酒、胡椒粉、芝麻油、水淀粉兑成芡汁。
3. 锅中放油烧至180℃，放入鱿鱼卷、姜片、马耳葱、蒜片爆炒出香，倒入芡汁，收汁亮油后装盘即成。

Ingredients
300g squid, 50g cooking oil

Seasonings
6g ginger (sliced), 10g garlic (sliced), 15g scallion (cut diagonally into sections shaped like horse ears), 2g salt, 20g sugar, 15g vinegar, 1g MSG, 10g Shaoxing cooking wine, 1g ground white pepper, 5g sesame oil, 10g cornstarch-water mixture

Preparation
1. Cut the squid into the shape of lichi flowers, and blanch in water.
2. Mix the salt, sugar, vinegar, MSG, Shaoxing cooking wine, ground white pepper, sesame oil and cornstarch-water mixture to make thickening sauce.
3. Heat oil in a wok to 180℃, add squid rolls, ginger, scallion and garlic, and stir-fry to bring out the aroma. Pour in the thickening sauce, wait till the sauce becomes thick and lustrous, and then transfer to a serving dish.

色泽红亮，麻辣鲜香
bright and lustrous colors; pungent, numbing and aromatic taste

香辣蟹
Hot-and-Spicy Crabs

[原　料]
螃蟹500克　食用油1000克（约耗30克）

[调　料]
干辣椒20克　花椒15克　香辣酱20克　豆豉10克　姜片10克　蒜片10克　葱丁15克　食盐3克　料酒20克　醪糟汁20克　味精1克　芝麻油5克　各种香料20克　干淀粉20克　火锅老油（食用油加辣椒、花椒、香料、姜、葱炒制而成）50克

[制　作]
1. 螃蟹初加工后加食盐、料酒、干淀粉拌匀，入180℃食用油中炸至色红、酥香时捞出。
2. 锅中放火锅老油烧至120℃，下干辣椒、花椒、豆豉、香辣酱、姜片、蒜片、葱丁及各种香料炒香，放入螃蟹、料酒、醪糟汁、味精、芝麻油炒入味，起锅装盘即成。

Ingredients
500g crabs, 1000g cooking oil for deep-frying

Seasonings
20g dried chilies, 15g Sichuan pepper, 20g chili pepper paste, 10g fermented soy beans, 10g ginger (sliced), 10g garlic (sliced), 15g scallion (finely chopped), 3g salt, 20g Shaoxing cooking wine, 20g fermented glutinous rice wine, 1g MSG, 5g sesame oil, 20g mixed herbal spices, 20g cornstarch, 50g hot pot oil (made by stir-frying chilies, Sichuan pepper, mixed herbal spices, ginger and scallion in cooking oil)

Preparation
1. Kill and wash the crabs thoroughly. Mix the crabs with salt, Shaoxing cooking wine and cornstarch, blend well and deep-fry in 180℃ oil till red, crisp and aromatic.
2. Heat the hot pot oil in a wok to 120℃, add dried chilies, Sichuan pepper, mixed herbal spices, fermented soy beans, chili pepper paste, ginger, scallion, garlic and stir-fry till aromatic. Blend in the crabs, Shaoxing cooking wine, fermented glutinous rice wine, MSG and sesame oil, and stir-fry till the crabs have absorbed the rich flavors. Transfer to a serving dish.

香辣蟹 Hot-and-Spicy Crabs

081

煳辣鲜贝 Hula-Flavor Scallops

色泽棕红，鲜贝滑嫩，咸鲜甜酸，具有干辣椒和花椒的香辣和香麻味
brownish color; tender and smooth scallop meat; salty, sour and sweet taste; strongly flavored by chilies and Sichuan pepper

煳辣鲜贝
Hula-Flavor Scallops

[原 料]
鲜贝250克 食用油60克

[调 料]
干辣椒节10克 花椒4克 姜片7克 蒜片7克 葱丁15克 食盐2克 料酒10克 酱油7克 醋10克 白糖15克 味精1克 鲜汤25克 水淀粉20克 蛋清淀粉40克

[制 作]
1．将食盐、料酒、酱油、醋、白糖、味精、鲜汤、水淀粉调成荔枝味芡汁。
2．鲜贝去筋，滤去多余水分，与食盐、料酒、蛋清淀粉拌匀。
3．锅中放食用油烧至140℃，下干辣椒节、花椒、鲜贝炒至断生，续下姜片、蒜片、葱丁炒香，倒入荔枝味芡汁，待收汁亮油时装盘即成。

Ingredients
250g scallops, 60g cooking oil

Seasonings
10g dried chilies (cut into sections), 4g Sichuan pepper, 7g ginger (sliced), 7g garlic (sliced), 15g scallion (finely chopped), 2g salt, 10g Shaoxing cooking wine, 7g soy sauce, 10g vinegar, 15g sugar, 1g MSG, 25g everyday stock, 20g cornstarch-water mixture, 40g mixture of egg white and cornstarch

Preparation
1. Mix salt, Shaoxing cooking wine, soy sauce, vinegar, sugar, MSG, stock and cornstarch-water mixture to make lichi-flavor thickening sauce.
2. Remove the tendon of the scallops, drain and mix well with salt, Shaoxing cooking wine and the mixture of egg white and cornstarch.
3. Heat oil in a wok to 140℃, add dried chilies, Sichuan pepper and scallops, and stir-fry till just cooked. Add the ginger, garlic and scallion, and stir-fry till aromatic. Pour in the lichi-flavor sauce, and wait till the sauce becomes thick and lustrous. Transfer to a serving dish.

银鳕鱼外酥里嫩，咸鲜香辣，山椒味浓
crispy fish skin and tender fish meat; aromatic and pungent flavor with fragrance of pickled tabasco peppers

竹烤银鳕鱼
Roasted Cod on a Bamboo Platter

[原 料]

银鳕鱼500克 甜椒200克 山东大葱100克

[调 料]

食盐3克 味精2克 日本味噌10克 料酒20克 火锅油40克 芝麻油5克 野山椒水500克 葱花15克

[制 作]

1. 银鳕鱼改成0.6厘米厚的大片，用食盐、味精、日本味噌、料酒腌制30分钟；甜椒用山椒水浸泡3个小时后捞出，切成约0.3厘米大小的丁；将山东大葱改刀成约16厘米长的节，在烤盘中码放整齐。
2. 银鳕鱼置于葱上，放入烤箱的烤盘上，用底火120℃、面火260℃烤约20分钟至主料外酥内嫩时取出放在竹笆上。
3. 锅置旺火上，放入火锅油烧至150℃，加入甜椒丁炒香，再加食盐、味精、芝麻油、山椒水调好味，起锅淋在银鳕鱼上即成。

Ingredients

500g cod, 200g red bell peppers, 100g Shandong scallion

Seasonings

3g salt, 2g MSG, 10g Japanese miso, 20g Shaoxing cooking wine, 40g hot pot oil, 5g sesame oil, 500g brine of pickled tabasco peppers, 15g scallion (chopped)

Preparation

1. Cut the cod into 0.6cm-thick fillets, and marinate in the mixture of salt, MSG, miso and Shaoxing cooking wine for 30 minutes. Marinate the red bell peppers in the brine of pickled tabasco peppers for 3 hours, remove and cut into 0.3cm³ cubes. Cut the Shandong scallion into 16cm-long sections and stack neatly on a roast plate.
2. Lay the cod onto the scallion and roast in an oven with a surface layer temperature of 120℃ and a bottom layer temperatuare of 260℃ for about 20 minutes till the skin becomes crispy and the meat tender. Remove the cod from the oven and put onto a bamboo tray.
3. Heat oil in a wok to 150℃, add red bell peppers and stir-fry till aromatic. Add salt, MSG, sesame oil and brine of pickled tabasco peppers, blend well and pour over the cod. Spinkle with chopped scallion and serve.

藿香鲈鱼

Ageratum-Flavored Perch

色泽红亮，咸鲜微辣、略带甜酸，藿香味浓郁
bright brown color; salty, sour, sweet and somewhat hot taste; strong ageratum fragrance

[原　料]

鲈鱼500克　藿香末25克　食用油1000克（约耗75克）

[调料A]

食盐0.5克　料酒10克　姜片7克　葱段10克

[调料B]

泡辣椒末25克　郫县豆瓣20克　泡姜末25克　蒜米25克　葱花25克

[调料C]

料酒10克　酱油2克　味精5克　白糖5克　醋10克　芝麻油3克　水淀粉25克　鲜汤400克

[制　作]

1. 鲈鱼宰杀后洗净，与调料A拌匀码味15分钟。
2. 锅置火上，加清水、食盐、料酒烧沸，放入鲈鱼煮至熟透后捞出装盘。
3. 锅置火上，入食用油烧至120℃，放入调料B炒香，再放入调料C和藿香末烧至汁浓后起锅，浇在鲈鱼上即成。

Ingredients

500g perch, 25g ageratum powder, 1000g cooking oil for deep-frying

Seasonings A

0.5g salt, 10g Shaoxing cooking wine, 7g ginger (sliced), 10g scallion (cut into sections)

Seasonings B

25g pickled chilies (finely chopped), 20g Pixian chili bean paste, 25g pickled ginger (finely chopped), 25g garlic (finely chopped), 25g scallion (finely chopped)

Seasonings C

10g Shaoxing cooking wine, 2g soy sauce, 5g MSG, 5g sugar, 10g vinegar, 3g sesame oil, 25g cornstarch-water mixture, 400g everyday stock

Preparation

1. Kill and rinse the perch. Marinate in Seasonings A for 15 minutes.
2. Heat a wok over a flame. Add water, salt and Shaoxing cooking wine, and bring to a boil. Slide in the perch, boil till cooked through, remove and transfer to a serving dish.
3. Heat oil in a wok to 120℃, add Seasonings B and stir-fry till aromatic. Add Seasonings C and ageratum powder, and simmer till the sauce thickens. Pour the sauce over the perch.

Speckled Hind Fish with Green and Red Peppers
双椒石斑鱼

鱼肉细嫩，咸鲜微辣
tender fish; salty, delicate and slightly hot taste

[原 料]
石斑鱼1尾（约500克） 青尖椒50克 红尖椒50克

[调 料]
姜片15克 葱段20克 食盐4克 料酒15克

[制 作]
1. 石斑鱼初加工后在鱼身两面各剞5刀，深度约1厘米，用姜片、葱段、食盐、料酒码味15分钟。
2. 青尖椒、红尖椒分别切成小颗粒。
3. 鱼入条盘，将青尖椒、红尖椒放在鱼上，入笼用旺火蒸10分钟取出即成。

Ingredients
1 speckled hind (about 500g), 50g green chili peppers, 50g red chili peppers

Seasonings
15g ginger (sliced), 20g scallion (cut into sections), 4g salt, 15g Shaoxing cooking wine

Preparation
1. Kill and clean the fish. Make five 1cm-deep cuts into both sides of the fish. Marinate the fish in the mixture of the ginger, scallion, salt and Shaoxing cooking wine for 15 minutes.
2. Chop the red and green chili peppers into grains.
3. Transfer the fish to an oval platter, top with green and red pepper grains, and steam over a high flame for 10 minutes.

冰糖燕窝

色泽晶莹，软润滑爽，清冽甘甜

Bird's Nest with Rock Sugar

pure white color; smooth, soft, sweet and refreshing flavor

[原　料]

水发燕窝250克　甜樱桃25克

[调　料]

冰糖250克　清水500克

[制　作]

1. 水发燕窝放在小盆内，用温水冲泡2次，沥干水分后备用。
2. 锅置微火上，放清水、冰糖煮至糖化汁粘时用净纱布滤去杂质，取净糖汁150克冲入燕窝，滗去糖汁，再将剩余的净糖汁冲入燕窝，入笼用旺火蒸5分钟后取出，放入樱桃即成。

Ingredients

250g water-soaked edible bird's nest, 25g candied cherries

Seasonings

250g rock sugar, 500g water

Preparation

1. Put the bird's nest into a small basin, pour in warm water and leave to soak for a while. Drain off the water, pour in fresh warm water and leave to soak for a second time. Remove and drain.
2. Heat a wok over a low flame, add water and rock sugar. Wait till the rock sugar melts and the soup becomes slightly sticky, then filter with cheesecloth to remove the dregs. Pour 150g of the filtered soup into a bowl with the bird's nest in it. Decant the soup out of the bowl and then pour in the rest of the filterd soup. Transfer the bowl into a steamer and steam over a high flame for 5 minutes. Put in the cherry.

Bird's Nest in Consomme
清汤燕菜

汤汁清澈，燕窝雪白
clear soup and snowy bird's nest

[原 料]
干燕窝25克 清汤1500克

[调 料]
食盐2克 味精1克

[制 作]
1. 将干燕窝摘洗干净，盛入大汤碗内，掺入沸水后加盖焖发至燕窝质软、涨透。
2. 锅中加清汤烧沸，放入燕窝稍焖后捞出，盛入汤钵中；锅中另换清汤烧沸，调入食盐、味精，倒入汤钵中即成。

Ingredients
25g edible bird's nest, 150g consomme

Seasonings
2g salt, 1g MSG

Preparation
1. Rinse the bird's nest and transfer to a large soup bowl. Pour in boiling water, cover and wait till the bird's nest is fully soaked and soft.
2. Pour some consomme into the wok, bring to a boil, add the bird's nest and cover. Wait for a few minutes and remove the bird's nest from the wok and transfer to a soup bowl. Throw away the stock in the wok and add the rest of the consomme into the wok. Bring to a boil, blend in salt and MSG, and then pour into the soup bowl.

色泽红亮，掌形完整，肉质粑糯，咸鲜醇厚
bright and lustrous color; glutinous and tender meat; salty, aromatic and delicate taste

一品牦牛掌
Deluxe Yak Paws

[原 料]
牦牛掌1对（约1500克）母鸡1只 猪肘500克 熟火腿250克 化猪油50克

[调 料]
食盐5克 酱油10克 料酒50克 味精4克 胡椒粉2克 冰糖色25克 姜30克 葱段40克 芝麻油10克 鸡汤2000克

[制 作]
1. 牦牛掌初加工后用清水反复氽煮几次，入锅煮约2小时至牦牛掌变软，用小刀将掌骨全部去掉，再入锅加鸡汤、姜、葱段、料酒煨煮3次，除去腥膻味。
2. 将牦牛掌横切数刀，以不切穿掌底为度，并包入纱布中；母鸡斩块；猪肘切块。
3. 锅中放化猪油烧至120℃，入姜、葱段炒香，掺入鸡汤，续下鸡块、猪肘、火腿、食盐、酱油、料酒、胡椒粉、冰糖色和牦牛掌包，用小火烧2～3小时至粑软。
4. 捞出葱白段入盘中垫底，牦牛掌摆放在葱上呈一"品"字形，锅中汁液加入味精、芝麻油收浓，浇淋在牦牛掌上即成。

Ingredients
2 yak paws (about 1500g), 1 hen, 500g pork knuckle, 250g precooked ham, 50g lard

Seasonings
5g salt, 10g soy sauce, 50g Shaoxing cooking wine, 4g MSG, 2g ground white pepper, 25g caramel color, 30g ginger, 40g scallion (cut into sections), 10g sesame oil, 2000g chicken stock

Preparation
1. Blanch the cleaned yak paws in water several times, boil for about 2 hours till they become soft, and de-bone. Put the paws into a pot, add some chicken stock, ginger, scallion and Shaoxing cooking wine, and bring over a low flame to a boil and simmer for about 40 minutes. Repeat the simmering process two more times to remove the unpleasant smell of the paws.
2. Make several cuts into the paws and wrap them up in cheesecloth. Chunk the hen and pork knuckle.
3. Heat oil in a wok to 120℃, add the rest of the ginger and scallion, and stir-fry till aromatic. Add chicken stock, chicken, pork knuckle, ham, salt, soy sauce, Shaoxing cooking wine, ground white pepper, caramel color and wrapped paws. Braise over a low flame for 2 to 3 hours till soft and tender.
4. Fish out the scallion white in the soup and lay neatly on a serving dish. Place the yak paws beautifully onto the scallion. Add MSG and sesame oil to the soup, simmer till the soup thickens, and pour over the paws.

汤清色雅，鸽蛋滑嫩，竹荪松脆，咸鲜味美
clear soup; tender eggs; crunchy and puffy mushrooms;
delicate and salty taste

竹荪鸽蛋
Pigeon Eggs with Veiled Lady Mushrooms

[原　料]

鸽蛋300克　干竹荪25克　豌豆苗50克　清汤1500克

[制　作]

1. 竹荪用清水涨发至透，切成约5厘米长的片，入锅中用沸水氽过，再用清汤喂味；豌豆苗摘取苞尖，洗净备用。
2. 鸽蛋破壳，入锅煮成荷包蛋，捞入汤碗中，加入竹荪，灌上清汤；再将豌豆苗烫熟，点缀在竹荪四周即成。

Ingredients

300g pigeon eggs, 25g dried veiled lady mushrooms, 50g pea vine sprouts, 1500g consomme

Preparation

1. Soak the dried veiled lady mushrooms in water till swollen. Cut into 5cm-long slices, blanch in water, and then marinate in the consomme. Rinse the pea vine sprouts.
2. Poach the pigeon eggs, transfer to a soup bowl, add the mushrooms and pour in some consomme. Blanch the pea vine sprouts and lay around the mushrooms to garnish.

成菜美观，鱼肉细嫩，清淡可口
exquisite and beautiful appearances; tender fish; delicate taste

Steamed Longsnout Catfish Surrounded by Flowers

清蒸百花江团
Steamed Longsnout Catfish Surrounded by Flowers

[原 料]
江团1尾 猪网油1张（约250克） 鱼糁200克 清汤1500克

[调 料]
味精2克 醋30克 化猪油5克 葱15克 姜30克 芝麻油10克 食盐5克 红、绿、黄、黑色植物原料适量 胡椒粉1.5克 料酒40克

[制 作]
1. 江团初加工后在鱼身两侧各剞5～6刀，深约1厘米左右，入沸水中汆一水捞出，放入鱼盘中，用食盐、料酒、胡椒粉、姜、葱码味15分钟。
2. 取小圆碟10个，抹上化猪油，将鱼糁舀入抹平，上面用各色植物原料牵摆成不同花形，上笼用小火蒸熟保温。
3. 将码入味的江团沥干水气，置于蒸盘内，盖上猪网油，掺入清汤、料酒，入笼用旺火蒸熟后取出，拣去猪网油、姜、葱不用，将江团轻轻滑入鱼盘内。
4. 锅置火上，放入清汤，再把蒸盘内的原汁滗入锅内烧沸，续下胡椒粉、食盐、味精搅匀，浇入盘内，最后将蒸熟的鱼糁花摆在江团周围即成。
5. 姜剁蓉，加食盐、醋、芝麻油调匀成毛姜醋味碟，同江团一起上桌。

Ingredients
1 Longsnout catfish, 250g pork caul fat, 200g fish mince, 1500g consomme

Seasonings
2g MSG, 30g vinegar, 5g lard, 15g scallion, 30g ginger, 10g sesame oil, 5g salt, 1.5g ground white pepper, 40g Shaoxing cooking wine, Vegetables (of red, green, black and yellow colors)

Preparation
1. Make five to six cuts (abourt 1cm deep) into each side of the fish, blanch the fish in water, transfer to a plate and marinate with 3g salt, 20g Shaoxing cooking wine, 0.5g ground white pepper, 15g ginger and scallion for 15 minutes.
2. Get ten saucers, smear with lard, put fish mince on to them and flatten. Carve and arrange the vegetables onto these saucers so that they look like flowers. Steam the saucers over a low flame till the ingredients on them are cooked through, and keep warm in the steamer.
3. Drain the marinated fish, transfer to a steaming plate, and cover with pork caul fat. Add 250g consomme and the Shaoxing cooking wine into the plate, steam over a high flame till the fish is cooked through. Remove from the steamer, discard the caul fat, ginger and scallion, and then transfer the fish to a serving dish.
4. Heat a wok over a flame, add consomme and the soup in the steaming plate, bring to a boil and then add ground white pepper, 1g salt and MSG to stir well. Pour the contents in the wok over the fish. Arrange the steamed fish mince and vegetables around the fish.
5. Chop finely the ginger, add 1g salt, vinegar and sesame oil, stir well and serve as the dipping sauce.

汤浓味鲜，裙边软糯

delicate and viscous sauce; soft and glutinous shell flounce

红烧裙边

Red-Braised Shell Rims of Chinese Turtle

[原 料]

甲鱼裙边300克 鸡腿肉250克 菜心150克 食用油1500克（约耗80克）

[调 料]

葱段15克 姜片10克 料酒25克 食盐4克 糖色15克 酱油10克 胡椒粉2克 芝麻油15克 水淀粉10克 鸡汤1000克

[制 作]

1. 甲鱼裙边初加工后切成条，在沸水中氽水2次去掉腥味；鸡腿肉切成块。
2. 锅中放食用油烧至180℃，入鸡块炸至表面水分干后捞出。
3. 锅置火上，放入鸡块、姜片、葱段、鸡汤、食盐、料酒、糖色、酱油、胡椒粉烧沸，去掉浮沫，放入裙边，用小火烧至裙边软熟后捞出定碗，入笼内保温。
4. 锅中放食用油烧热，下菜心、食盐炒熟后装盘垫底，将裙边翻扣在菜心上；锅中放入烧裙边的原汁、芝麻油，用水淀粉勾芡，浇淋在裙边上即成。

Ingredients

300g shell rims of Chinese turtle, 250g chicken leg meat, 150g tender leafy vegetables, 1500g cooking oil for deep-frying

Seasonings

15g scallion (cut into sections), 10g ginger (sliced), 25g Shaoxing cooking wine, 4g salt, 15g caramel color, 10g soy sauce, 2g ground white pepper, 15g sesame oil, 10g cornstarch-water mixture, 1000g chicken stock

Preparation

1. Cut the shell rims into slivers, and blanch in boiling water twice to remove the unpleasant smell. Cut the chicken leg meat into chunks.
2. Heat oil in a wok to 180℃, deep-fry the chicken chunks till their outer surfaces are dry, and then remove from the wok.
3. Heat a wok over a flame, add the chicken chunks, ginger, scallion, chicken stock, salt, Shaoxing cooking wine, caramel color, soy sauce and ground white pepper, bring to a boil and skim. Add the shell rims and simmer till soft and cooked through. Stack the shell rims into a steaming bowl and transfer to a steamer to keep them warm.
4. Heat oil in a wok, add tender leafy vegetables and salt, stir-fry till fully cooked and transfer to a serving dish. Turn the steaming bowl upside down onto the serving dish so that the shell rim slivers cover the tender leafy vegetables. Heat the pre-simmered stock in a wok and stir in the sesame oil. Add the cornstarch-water mixture to thicken the sauce. Pour the sauce over the shell rims.

土豆烧甲鱼
Braised Chinese Turtle with Potatoes

色泽红亮，质地软糯，麻辣香醇
brown color; tender and glutinous turtle meat; spicy and aromatic taste

[原　料]

甲鱼500克　土豆400克　食用油800克（约耗120克）

[调　料]

郫县豆瓣40克　香辣酱20克　姜片10克　葱段30克　食盐2克　味精6克　胡椒粉1克　醪糟汁15克　料酒40克　辣椒油20克　花椒油5克　芝麻油5克　鲜汤300克

[制　作]

1. 甲鱼宰杀后入沸水略煮后捞出洗净，将甲鱼从背甲与底板处分开，去内脏后洗净，斩成3厘米见方的块，入160℃食用油中略炸后捞出。
2. 土豆去皮洗净，切成边长约3厘米的菱形块，入180℃食用油中炸至表皮变硬时捞出。
3. 锅置火上，入食用油烧至120℃，放入郫县豆瓣、香辣酱、姜片、葱段炒香，入鲜汤烧沸，去掉料渣，放甲鱼块、食盐、味精、胡椒粉、醪糟汁、料酒，用中小火烧至软熟，再放土豆烧至入味，加辣椒油、花椒油、芝麻油推匀，起锅入盘即成。

Ingredients

500g Chinese turtle, 400g potatoes, 800g cooking oil for deep-frying

Seasonings

40g Pixian chili bean paste, 20g chili pepper paste, 10g ginger (sliced), 30g scallion (cut into sections), 2g salt, 6g MSG, 1g ground white pepper, 15g fermented glutinous rice wine, 40g Shaoxing cooking wine, 20g chili oil, 5g Sichuan pepper oil, 5g sesame oil, 300g everyday stock

Preparation

1. Kill the turtle. Boil the turtle briefly in water, remove, rinse, and separate back shell from bottom shell. Remove the entrails, rinse, chop into 3cm³ cubes and fry briefly in 160℃ oil.
2. Peel the potatoes, rinse, chop into 3cm-long diamonds and fry in 180℃ oil till the surface hardens.
3. Heat some oil in a wok to 120℃, add Pixian chili bean paste, chili pepper paste, ginger, scallion and stir-fry till aromatic. Add the stock and bring to a boil. Remove the scums, and slide in the turtle cubes. Add the salt, MSG, ground white pepper, fermented glutinous rice wine, Shaoxing cooking wine and simmer over a medium-low flame till the turtle cubes become soft and cooked through. Add the potatoes and braise still they have absorbed the flavors. Blend in chili oil, Sichuan pepper oil and sesame oil, mix well, transfer to a serving dish.

川式烤鳗鱼
Sichuan-Style Barbecued Eel

pleasantly bright color; tender eel; hot, spicy and slightly sweet taste

色泽美观，肉质细嫩，麻辣微甜

[原 料]
鳗鱼500克

[调 料]
泡辣椒末30克 辣椒粉5克 花椒粉2克 饴糖30克 酱油10克 姜片5克 料酒10克 白芝麻5克 蜂蜜10克

[制 作]
1. 鳗鱼去头、尾、骨刺后洗净，切段，加饴糖、料酒、姜片、酱油浸泡20分钟；取出鳗鱼，剩余汤汁加泡辣椒末、辣椒粉、花椒粉调成酱汁。
2. 用竹签串上鳗鱼肉，用230℃温度烤制10分钟，中途不时翻面并涂抹酱汁，取出鱼段，刷上蜂蜜，撒上白芝麻即成。

Ingredients
500g eel

Seasonings
30g pickled red chilies (minced), 5g ground chili, 2g ground roasted Sichuan pepper, 30g maltose syrup, 10g soy sauce, 5g ginger (sliced), 10g Shaoxing cooking wine, 5g white sesame seeds, 10g honey

Preparation
1. Remove the head, caudal fin, bones of the eel, rinse in water, and then cut into chunks. Marinate the chunks in the mixture of the maltose syrup, Shaoxing cooking wine, ginger slices and soy sauce for 20 minutes. Remove the eel from the marinade, and add minced pickled chilies, ground chilies and ground roasted Sichuan pepper to the marinade to make the seasoning sauce.
2. Skewer the eel with bamboo sticks and grill over a 230℃ fire for 10 minutes. Turn the eel constantly and smear with the seasoning sauce. Remove the eel from the sticks, coat with honey and sprinkle with the white sesame seeds.

Fish in Chili Bean Sauce
豆瓣鱼

色泽红亮，鱼肉细嫩，味咸鲜微辣，略带酸甜

reddish color; tender fish; salty, spicy and slightly sour and sweet taste

[原　料]

鲜鱼1尾（约600克）食用油1000克（约耗100克）

[调料A]

食盐1克　姜片7克　葱段15克　料酒10克

[调料B]

郫县豆瓣35克　姜米10克　蒜米15克　葱花25克　白糖10克　酱油8克　醋10克　鸡精2克　料酒10克　鲜汤500克　水淀粉25克

[制　作]

1. 鲜鱼初加工后在鱼身两面各剞数刀，用调料A拌匀码味15分钟。
2. 锅中下食用油烧至210℃，入鱼炸至皮紧、色浅黄时捞出。
3. 锅中放食用油烧至120℃，放入郫县豆瓣、姜米、蒜米、葱花炒出香，掺入鲜汤，下鱼、白糖、酱油、料酒、醋、鸡精，烧至鱼软、成熟时捞出，锅中汁液用水淀粉收浓，起锅淋在鱼上即成。

Ingredients

1 fish (about 600g), 1000g cooking oil for deep-frying

Seasonings A

1g salt, 7g ginger (sliced), 15g scallion (cut into sections), 10g Shaoxing cooking wine

Seasonings B

35g Pixian chili bean paste, 10g ginger (finely chopped), 15g garlic (finely chopped), 25g scallion (finely chopped), 10g sugar, 8g soy sauce, 10g vinegar, 2g chicken essence granules, 10g Shaoxing cooking wine, 500g everyday stock, 25g cornstarch-water mixture

Preparation

1. Kill and clean the fish thoroughly. Make a couple of cuts into each side of the fish. Marinate the fish in the sauce made by mixing Seasonings A.
2. Heat oil in a wok to 210℃, fry the fish till the skin wrinkles and browns, then remove.
3. Heat some oil in the wok to 120℃, add the chili bean paste, ginger, garlic, scallion, and stir-fry till aromatic. Add the stock, fish, sugar, soy sauce, Shaoxing cooking wine, vinegar and chicken essence granules, and braise till the fish is soft and cooked through. Remove the fish. Thicken the sauce in the wok with cornstarch-water mixture and pour over the fish.

鱼嫩汤鲜
tender and delicate ingredients

砂锅雅鱼
Ya Fish Casserole

[原　料]

雅鱼500克　熟鸡肉30克　熟猪肚30克　熟猪舌30克　熟猪心30克　熟火腿40克　水发鱿鱼80克　水发香菇30克　冬笋30克　豆腐150克　奶汤1000克

[调　料]

姜片5克　葱段10克　蒜片5克　食盐5克　味精2克　胡椒粉1克　料酒30克　鸡油30克

[制　作]

1. 鸡肉、猪舌、猪心、猪肚、火腿、冬笋、香菇切成片；豆腐切成长约4厘米、粗约1.5厘米的条。
2. 将上述原料连同姜片、葱段、蒜片、食盐、胡椒粉、料酒放入砂锅内，再放雅鱼、奶汤置旺火上烧沸，改用小火煮20分钟，加鱿鱼片、味精、鸡油煮5分钟即成。

Ingredients

500g Ya Fish (Schizothorax prenanti), 30g pre-cooked chicken, 30g pre-cooked pork tripe, 30g pre-cooked pork tongue, 30g pre-cooked pork heart, 40g pre-cooked ham, 80g water-soaked squid, 30g water-soaked shiitake mushrooms, 30g winter bamboo shoots, 150g tofu, 1000g milky soup

Seasonings

5g ginger (sliced), 10g scallion (cut into sections), 5g garlic (sliced), 5g salt, 2g MSG, 1g ground white pepper, 30g Shaoxing cooking wine, 30g chicken oil

Preparation

1. Slice the chicken, pork tongue, heart, tripe, ham, bamboo shoots and shiitake mushrooms. Cut tofu into strips about 4cm long and 1.5cm thick.
2. Blend the ingredients mentioned in step 1 with the ginger, scallion, garlic, salt, ground white pepper and Shaoxing cooking wine in an earthen casserole. Add the fish and milky soup, bring to a boil over a high flame and simmer over a low flame for 20 minutes. Add the squid, MSG and chicken oil, and continue to boil for another five minutes.

香辣黄蜡丁
Hot-and-Spicy Yellow Catfish

鱼肉质地细嫩，口感麻、辣、咸、鲜、烫，香味浓郁

tender fish; aromatic smell; a combination of numbing, hot, salty, pungent and savoury taste

香辣黄蜡丁
Hot-and-Spicy Yellow Catfish

[原 料]
黄蜡丁500克 蒜苗50克 芹菜75克 青笋尖100克 香菜10克 食用油100克

[调料A]
刀口辣椒（干辣椒、花椒炒香，剁碎）20克 葱花10克

[调料B]
豆瓣20克 泡辣椒末15克 姜米10克 蒜米10克

[调料C]
食盐4克 酱油10克 水淀粉20克 醋3克 料酒15克 胡椒粉1克 味精2克 鲜汤700克

[制 作]
1. 蒜苗拍破，与芹菜同切成约10厘米长的段；青笋尖切成约10厘米长的薄片；将三料混合入锅，加少许食用油和食盐炒断生，捞出后放入汤碗垫底。
2. 锅置火上，下食用油烧至120℃，放入调料B炒香，续入调料C和黄蜡丁，用小火烧至黄蜡丁软熟入味，用水淀粉勾芡，盛入碗中，加调料A，淋上200℃的热油少许，放上香菜即成。

Ingredients
500g yellow catfish, 50g leeks, 75g celery, 100g asparagus lettuce tips (the leafy part), 10g coriander leaves, 100g cooking oil

Seasonings A
20g blade-minced chilies (stir-fry dried chilies and Sichuan pepper, and then finely chopped), 10g scallion (finely chopped)

Seasonings B
20g chili bean paste, 15g pickled chilies (finely chopped), 10g ginger (finely chopped), 10g garlic (finely chopped)

Seasonings C
4g salt, 10g soy sauce, 20g cornstarch-water mixture, 3g vinegar, 15g Shaoxing cooking wine, 1g ground white pepper, 2g MSG, 700g everyday stock

Preparation
1. Smash the leeks with the flat side of a blade, and then cut into 10cm-long sections. Cut the celery into 10cm lengths and the asparagus lettuce tips into 10cm-long slices. Stir-fry the three vegetables till al dente and transfer to a soup bowl.
2. Heat oil in a wok to 120℃, add Seasonings B, and stir-fry till aromatic. Add Seasonings C and yellow catfish, and braise over a low flame till the fish is soft and cooked through. Add the cornstarch-water mixture to thicken the sauce, transfer to the bowl, add Seasonings A and pour some 200℃ oil over the fish. Sprinkle with coriander leaves.

Mala Crayfish
麻辣小龙虾

色泽红亮，咸鲜麻辣
bright color; salty, pugent, numbing and appetizing taste

[原 料]

小龙虾500克 香芹节150克 食用油1000克（约耗150克）

[调 料]

干辣椒200克 花椒50克 大蒜100克 姜片15克 葱段20克 八角5克 香叶4克 草果5克 火锅底料50克 食盐2克 料酒20克 味精5克 胡椒粉2克 芝麻油5克 鲜汤100克

[制 作]

1. 小龙虾初加工后入160℃的食用油中炸1分钟捞出。
2. 锅中放食用油烧至120℃，入干辣椒、花椒、大蒜、姜片、葱段、八角、香叶、草果、火锅底料炒香，再入小龙虾、鲜汤、食盐、料酒、胡椒粉，用中火烧5分钟，待水分将干时放入香芹、味精、芝麻油炒匀装盘即成。

Ingredients

500g crayfish, 150g celery (cut into sections), 1000g cooking oil for deep-frying

Seasonings

200g dried chilies, 50g Sichuan pepper, 100g garlic, 15g ginger (sliced), 20g scallion (cut into sections), 5g star anise, 4g bay leaves, 5g tsaoko amomum, 50g hot pot seasonings, 2g salt, 20g Shaoxing cooking wine, 5g MSG, 2g ground white pepper, 5g sesame oil, 100g everyday stock

Preparatiaon

1. Purge the crayfish in water so as to clear the digestive tract. Kill, clean and deep-fry in 160℃ oil for 1 minute.
2. Heat oil in a wok to 120℃, add the dried chilies, Sichuan pepper, garlic, ginger, scallion, star anise, bay leaves, tsaoko amomum and hot pot seasoning, and then stir-fry to bring out the aroma. Add the stock, salt, Shaoxing cooking wine and ground white pepper. Simmer over a medium flame for 5 minutes. Add the celery, MSG and sesame oil when the sauce thickens, blend well and transfer to a serving dish.

色泽红亮，肉质细嫩，咸鲜微辣
bright reddish color; tender meat; salty, savory and slightly hot taste; lingering fragrance

泡椒牛蛙

Pickled-Chili-Flavored Bullfrog

[原　料]

牛蛙450克　泡红辣椒100克　食用油800克（约耗60克）

[调　料]

姜片10克　大蒜30克　葱段20克　食盐4克　味精2克　胡椒粉1克　醪糟汁15克　料酒40克　花椒油5克　芝麻油5克　鲜汤300克　水淀粉15克　泡辣椒油80克

[制　作]

1. 牛蛙宰杀后斩成3厘米见方的块，入150℃的食用油中略炸后捞出。
2. 锅置火上，下泡辣椒油40克烧至120℃，放入泡红辣椒、姜片、大蒜、葱段炒香，加鲜汤、牛蛙、食盐、味精、胡椒粉、醪糟汁、料酒烧至熟软，再加泡辣椒油、花椒油、芝麻油，用水淀粉勾芡，起锅装盘即成。

Ingredients

450g bullfrogs, 100g pickled red chilies, 800g cooking oil for deep-frying

Seasonings

10g ginger (sliced), 30g garlic, 20g scallion (cut into sections), 4g salt, 2g MSG, 1g ground white pepper, 15g fermented glutinous rice wine, 40g Shaoxing cooking wine, 5g Sichuan pepper oil, 5g sesame oil, 300g everyday stock, 15g cornstarch-water mixture, 80g pickled chili oil

Preparation

1. Kill and rinse the bullfrogs, chop into (3cm)³ cubes and deep-fry briefly in 150℃ oil.
2. Put a wok over a flame, add 40g pickled chili oil and bring to 120℃. Add pickled red chilies, ginger, garlic, scallion, stock, bullfrogs, salt, MSG, ground white pepper, fermented glutinous rice wine and Shaoxing cooking wine, then simmer till the bullfrog cubes are soft and cooked through. Add pickled chili oil, Sichuan pepper oil, and sesame oil. Thicken with the cornstarch-water mixture, remove from the heat and transfer to a serving dish.

开门红

Good-Luck Fish Head

色泽红亮，鱼头细嫩，鲜辣不燥
bright red color; tender fish head; spicy and delicate taste

[原　料]
花鲢鱼头1个（约800克）　红甜椒300克

[调料A]
食盐2克 料酒10克 姜片7克 葱段10克 花椒1克

[调料B]
青尖椒粒50克 小米红椒粒100克 野山椒粒50克 食盐4克 胡椒粉2克 味精2克 料酒20克 姜片10克 葱段15克 老抽10克 食用油50克

[制　作]
1. 鱼头洗净后对剖成相连的两半，用调料A码味15分钟；红甜椒对剖成大片；调料B调匀成味汁。
2. 将鱼头放入大汤盘中淋上味汁，盖上红甜椒片，入笼内用旺火蒸12分钟至熟后取出即成。

Ingredients
head of one silver carp (about 800g), 300g red bell peppers

Seasonings A
2g salt, 10g Shaoxing cooking wine, 7g (sliced), 10g scallion (cut into sections), 1g Sichuan pepper

Seasonings B
50g green chili peppers (finely chopped), 100g birdseye chilies (finely chopped), 50g Tabasco peppers (finely chopped), 4g salt, 2g ground white pepper, 2g MSG, 20g Shaoxing cooking wine, 10g ginger (sliced), 15g scallion (cut into sections), 10g soy sauce, 50g cooking oil

Preparation
1. Rinse the fish head, halve and marinate in Seasonings A for 15 minutes. Halve the red bell peppers. Mix Seasonings B to make the seasoning sauce.
2. Lay the fish head on a large soup dish, drizzle with the sauce, cover with the red bell peppers and steam over a high flame for 12 minutes.

Bullfrog in a Stone Pot
石锅牛蛙

肉质细嫩，咸鲜微辣
tender bullfrog; salty, savoury and slightly hot taste

[原料]

牛蛙400克 大蒜80克 食用油1000克（约耗150克）

[调料]

郫县豆瓣30克 泡辣椒段25克 姜米10克 葱段20克 食盐2克 料酒40克 味精3克 鲜汤300克

[制作]

1. 牛蛙初加工后斩成约3厘米见方的块，加食盐、料酒拌匀，放入150℃的食用油中过油后捞出。
2. 锅内放食用油、大蒜不断翻炒，再加郫县豆瓣、泡辣椒段、姜米、葱段炒香，续下食盐、料酒、味精、鲜汤、牛蛙烧至软熟，起锅后装入烧热的石锅内即成。

Ingredients

400g bullfrogs, 80g garlic, 1000g cooking oil for deep-frying,

Seasonings

30g Pixian chili bean paste, 25g pickled chilies (cut into sections), 10g ginger (finely chopped), 20g scallion (cut into sections), 2g salt, 40g Shaoxing cooking wine, 3g MSG, 300g everyday stock

Preparation

1. Kill and clean the bullfrogs. Cut them into 3cm^3 cubes, add salt and Shaoxing cooking wine, mix well and then deep-fry briefly in 150℃ oil. Remove.
2. Heat some oil in a wok, add the garlic and stir-fry. Blend in Pixian chili bean paste, pickled chilies, ginger, scallion, salt, Shaoxing cooking wine, MSG, stock and bullfrogs, braise till the bullfrog cubes are soft and cooked through. Transfer to a stone pot.

川味烤鱼
Sichuan-Flavor Barbecued Fish

味透入骨,鲜香浓郁

full absorption of the flavors into the fish; pungent and aromatic taste

川味烤鱼
Sichuan-Flavor Barbecued Fish

[原 料]

草鱼750克 黄豆芽50克 青笋条50克 食用油100克

[调 料]

姜片10克 蒜片20克 葱花20克 泡辣椒30克 郫县豆瓣20克 香辣酱20克 老干妈豆豉10克 蚝油10克 香料粉5克 食盐5克 味精4克 胡椒粉2克 料酒50克

[制 作]

1. 草鱼从背部开膛,去内脏后洗净,用食盐、料酒腌制片刻,放炭火架上烤至九成熟。
2. 锅置火上,加食用油烧热,下姜片、蒜片、泡辣椒、郫县豆瓣、香辣酱、老干妈豆豉、蚝油、香料粉、味精、胡椒粉、料酒炒匀成酱料。
3. 在玻璃方盘内铺上黄豆芽和青笋条,鱼放其上,浇上炒好的酱料,撒上葱花。
4. 玻璃方盘底用火加热保温即成。

Ingredients

750g grass carp, 50g soybean sprouts, 50g asparagus lettuce (cut into slivers), 100g cooking oil

Seasonings

10g ginger (sliced), 20g garlic (sliced), 20g scallion (finely chopped), 30g pickled chilies, 20g Pixian chili bean paste, 20g chili pepper paste, 10g Laoganma fermented soy beans, 10g oyster sauce, 5g mixed herbal spices (ground), 5g salt, 4g MSG, 2g ground white pepper, 50g Shaoxing cooking wine

Preparation

1. Kill the fish by cutting into the back, gut it and wash thoroughly. Marinate the fish with salt and Shaoxing cooking wine. Put the fish on the grill and roast till just cooked.
2. Heat oil in a wok, add the ginger, garlic, pickled chilies, Pixian chili bean paste, chili pepper paste, Laoganma fermented soy beans, oyster sauce, herbal spice, MSG, ground white pepper and Shaoxing cooking wine, and blend well to make seasoning sauce.
3. Put the soybean sprouts and asparagus lettuce slivers into a large square glass plate. Lay the fish on top of them, pour over the seasoning sauce, and sprinkle with scallion.
4. Keep the glass plate warm by heating over a flame.

色泽红亮，肉质鲜嫩，咸鲜味浓
deep red color; tender and soft fish;
salty and savory taste

干烧鱼
Dry-Braised Fish

[原 料]
草鱼1尾 猪肉臊70克 食用油1500克（约耗50克） 鲜汤500克

[调料A]
食盐2克 姜片10克 葱段15克 料酒10克

[调料B]
芽菜末10克 泡椒段10克 姜米8克 蒜米10克 葱段50克

[调料C]
食盐2克 料酒15克 醪糟汁30克 酱油10克 味精1克 芝麻油10克

[制 作]
1. 草鱼宰杀后去内脏洗净，在鱼身两侧各剞数刀，用调料A码味15分钟，入200℃的食用油中炸至皮紧、色呈浅棕红时捞出。
2. 锅置火上，入食用油烧至120℃，放调料B炒香，掺入鲜汤，下草鱼、猪肉臊、调料C烧沸，改用中小火烧至亮油后起锅装盘即成。

Ingredients
1 grass carp, 70g stir-fried pork mince, 1500g cooking oil for deep-frying, 500g everyday stock

Seasonings A
2g salt, 10g ginger (sliced), 15g scallion (cut into sections), 10g Shaoxing cooking wine

Seasonings B
10g minced yacai (preserved mustard stems), 10g pickled chilies (segmented), 8g ginger (finely chopped), 10g garlic (finely chopped), 50g scallion (cut into sections)

Seasonings C
2g salt, 15g Shaoxing cooking wine, 30g fermented glutinous rice wine, 10g soy sauce, 1g MSG, 10g sesame oil

Preparation
1. Kill, clean and rinse the fish. Make several cuts in each side and marinate in Seasonings A for 15 minutes. Deep-fry the fish in 200˚C oil till the skin tightens and becomes light brown. Remove.
2. Heat oil in a wok to 120˚C, add Seasonings B, and stir-fry till aromatic. Add the stock, fish and Seasonings C, bring to a boil and then simmer over a medium-low flame till the sauce becomes lustrous. Remove from the stove and transfer to a serving dish.

色泽金黄，外酥内嫩，甜酸味浓，造型美观

exquisite form; tender fish that is crispy on the outside; sweet and sour taste; golden brown color

糖醋脆皮鱼
Crispy Sweet-and-Sour Fish

[原　料]

草鱼1条（约750克）　食用油2000克（约耗80克）

[调料A]

姜片5克　葱段10克　食盐1克　料酒10克

[调料B]

泡辣椒丝10克　葱丝10克　姜米6克　蒜米10克　葱花15克　食盐1克　白糖30克　醋30克　酱油6克　鲜汤200克　水淀粉100克

[制　作]

1. 草鱼初加工后在鱼身两侧各刻5刀，用调料A码味15分钟，再用水淀粉挂糊，放入180℃的食用油中炸定型后捞出，待油温回升到200℃时，再将鱼放入炸至色黄、酥脆时捞出装入盘中。
2. 炒锅置火上，下少许食用油烧至120℃，加入姜米、蒜米、葱花炒香，再下鲜汤、食盐、醋、酱油、白糖、水淀粉收汁，起锅浇在鱼上，撒上泡辣椒丝、葱丝即成。

Ingredients

1 grass carp (about 750g), 2000g cooking oil for deep-frying

Seasonings A

5g ginger (sliced), 10g scallion (cut into sections), 1g salt, 10g Shaoxing cooking wine

Seasonings B

10g pickled chilies (cut into strips), 10g scallion (cut into strips), 6g ginger (finely chopped), 10g garlic (finely chopped), 15g scallion (finely chopped), 1g salt, 30g sugar, 30g vinegar, 6g soy sauce, 200g everyday stock, 100g cornstarch-water mixture

Preparation

1. Kill the fish and wash thoroughly. Make five cuts in eah side of the fish and marinate in Seasonings A for 15 minutes. Coat with the cornstarch-water mixture, deep-fry in 180℃ oil for one or two minutes, and remove. Heat the oil to 200℃, deep-fry the fish for a second time till the skin becomes brown and crispy, and transfer to a serving dish.
2. Heat some oil in a wok to 120℃, add chopped ginger, garlic and scallion, and stir-fry to bring out the fragrance. Add the stock, salt, vinegar, soy sauce, sugar and cornstarch-water mixture. Wait till the sauce thickens and pour it over the fish. Sprinkle with pickled chilies and scallion strips.

芹黄熘鱼丝
Stir-Fried Fish Slivers with Celery

滑嫩清香，咸鲜可口
tender and smooth fish; salty and savoury taste

[原 料]
净鱼肉200克 芹黄100克 食用油1000克（约耗100克）

[调 料]
泡辣椒丝5克 姜丝5克 蒜丝10克 食盐2克 味精1克 胡椒粉1克 醋3克 鲜汤50克 水淀粉10克 蛋清淀粉35克 芝麻油5克 料酒5克

[制 作]
1. 芹黄洗净后切成约3.5厘米长的节；鱼肉切为二粗丝，装入碗中，用食盐、料酒、蛋清淀粉拌匀待用。
2. 食盐、味精、胡椒粉、醋、芝麻油、鲜汤、水淀粉调成芡汁。
3. 炒锅置火上，下食用油烧至80℃，入鱼丝滑散籽，滗去余油，放入泡辣椒丝、姜丝、蒜丝、芹黄节炒出香味，倒入芡汁，收汁亮油后装盘即成。

Ingredients
200g fish, 100g tender stems of celery, 1000g cooking oil for deep-frying

Seasonings
5g pickled chilies (shredded), 5g ginger (shredded), 20g garlic (shredded), 2g salt, 1g MSG, 1g ground white pepper, 3g vinegar, 50g everyday stock, 10g cornstarch-water mixture, 35g mixture of egg white and cornstarch, 5g sesame oil, 5g Shaoxing cooking wine

Preparation
1. Rinse the celery and cut into 3.5cm-long sections. Cut the fish into slivers about 8cm long and 0.3cm thick, transfer to a bowl and mix well with salt, Shaoxing cooking wine and the mixture of egg white and cornstarch.
2. Mix salt, MSG, ground white pepper, vinegar, sesame oil, stock and cornstarch-water mixture to make thickening sauce.
3. Heat oil in a wok to 80℃, add fish slivers and deep-fry, stirring so that they separate. Drain off the excess oil, add pickled chilies, ginger, garlic and celery, and stir-fry to bring out the aroma. Pour in the thickening sauce, and wait till the sauce is thick and lustrous. Transfer to a serving dish.

酸菜鱼
Fish with Pickled Mustard

鱼片洁白，质地细嫩，咸鲜酸辣
tender fish with salty, sour and hot taste

[原　料]
草鱼肉500克　泡青菜250克　食用油80克

[调　料]
野山椒50克　姜片10克　葱段15克　鸡蛋清20克　干淀粉20克　鲜汤1000克　食盐1克　料酒15克　鸡精2克　胡椒粉2克　葱花7克　蒜泥15克

[制　作]
1. 草鱼肉片成厚片，加入食盐、料酒、鸡蛋清、干淀粉拌匀。
2. 锅中放食用油烧至140℃，下泡青菜、野山椒、姜片、葱段炒香，掺入鲜汤，放入料酒、胡椒粉、鸡精烧沸出味，续下鱼片煮熟，倒出装入大汤碗内，撒上葱花、蒜泥，淋少许热油即成。

Ingredients
500g grass carp meat, 250g pickled mustard greens, 80g cooking oil

Seasonings
50g tabasco pepperg, 10g ginger (sliced), 15g scallion (cut into sections), 20g egg white, 20g cornstarch, 1000g everyday stock, 1g salt, 15g Shaoxing cooking wine, 2g chicken essence granules, 2g ground white pepper, 7g scallion (finely chopped), 15g garlic (finely chopped)

Preparation
1. Cut the fish into thick fillets, mix with salt, Shaoxing cooking wine, egg white and cornstarch, and blend well.
2. Heat oil in a wok to 140℃, add pickled mustard, tabasco pepperg, ginger and scallion to stir-fry till aromatic. Blend in the stock, Shaoxing cooking wine, ground white pepper and chicken essence granules, bring to a boil, and slide in the fillets. Boil till the fish is cooked through. Pour the contents in the wok into a large soup bowl, sprinkle with scallion and garlic, and drizzle with some hot oil.

色泽红亮，肉质细嫩，咸鲜酸甜微辣
bright brown color; tender catfish; rich and slightly hot taste

软烧仔鲶
Braised Catfish

[原　料]

仔鲶鱼500克　独蒜100克　食用油1500克（约耗150克）

[调　料]

郫县豆瓣30克　泡辣椒末15克　姜米10克　葱花20克　食盐2克　酱油10克　醋20克　料酒40克　白糖25克　味精3克　水淀粉30克　鲜汤500克

[制　作]

1． 鲶鱼宰杀后在鱼背处宰2~3刀，加食盐、料酒码味。
2． 锅中放食用油烧热，放入独蒜炒香，加郫县豆瓣、泡辣椒末、姜米炒出香味，续下鲶鱼炒至表面变色，放入鲜汤、食盐、酱油、料酒、白糖，用小火烧至软熟入味，最后下葱花、味精、醋、水淀粉，收汁后装盘即成。

Ingredients
500g catfish, 100g garlic, 1500g cooking oil

Seasonings
30g Pixian chili bean paste, 15g pickled chilies (finely chopped), 10g ginger (finely choppedd), 20g scallion (finely chopped), 2g salt, 10g soy sauce, 20g vinegar, 40g Shaoxing cooking wine, 25g sugar, 3g MSG, 30g cornstarch-water mixture, 500g everyday stock

Preparation
1. Kill and core the catfish. Make two or three cuts into the back of the fish. Add salt and Shaoxing cooking wine to marinate.
2. Heat oil in a wok and stir-fry the garlic till aromatic. Add the Pixian chili bean paste, pickled chilies and ginger, stir-fry till aromatic, and add the catfish. Continue to stir-fry till the color of the fish skin changes. Add the stock, salt, soy sauce, Shaoxing cooking wine and sugar, simmer till the fish is soft and cooked through, and blend in the chopped scallion, MSG, vinegar and cornstarch-water mixture to thicken the sauce. Transfer to a plate.

鱼片细嫩，豆芽清香，香味浓郁
tender fish; fragrant soy bean sprouts; lingering and aromatic smell

香辣沸腾鱼
Hot-and-Spicy Sizzling Fish

[原　料]

草鱼1尾（约1000克）　黄豆芽100克　食用油600克　鲜汤1000克

[调料A]

食盐1克　料酒10克　鸡精2克　蛋清淀粉10克

[调料B]

食盐3克　姜片5克　葱段10克　料酒15克　鸡精2克　味精1克　胡椒粉2克

[调料C]

干红辣椒100克　花椒40克　香辣酱10克

[制　作]

1. 草鱼初加工后取净鱼肉片成约0.3厘米厚的片，与调料A拌匀；鱼头、鱼骨斩成小块。
2. 黄豆芽入锅炒熟，装入大汤钵内垫底。
3. 锅中掺鲜汤，下调料B烧沸，放入鱼头、鱼骨煮熟后捞出，放在黄豆芽上；再将鱼片入锅煮至半熟后捞出，放在鱼骨上。
4. 锅中放食用油烧至120℃，放入调料C炒香，起锅淋在鱼片上即成。

Ingredients

1 grass carp (about 1000g), 100g soy bean sprouts, 600g cooking oil, 1000g everyday stock

Seasonings A

1g salt, 10g Shaoxing cooking wine, 2g chicken essence granules, 10g mixture of egg white and cornstarch

Seasonings B

3g salt, 5g ginger (sliced), 10g scallion (cut into sections), 15g Shaoxing cooking wine, 2g chicken essence granules, 1g MSG, 2g ground white pepper

Seasonings C

100g dried chilies, 40g Sichuan pepper, 10g chili pepper paste

Preparation

1. Clean the fish, cut its meat into 0.3cm-thick slices, mix with Seasonings A and blend well. Finely chop the fish head and bones.
2. Stir-fry the soy bean sprouts and transfer to a large soup bowl.
3. Heat the stock in a wok, add Seasonings B, and bring to a boil. Add chopped fish head and bones, boil till the fish head and bones are cooked through, remove and lay on top of the soy bean sprouts. Add fish slices into the wok, boil till just cooked, remove and lay on top of the bones.
4. Heat oil in a wok to 120℃, add Seasonings C, stir-fry till fragrant and pour over the fish slices.

色泽红亮，质感滑嫩，咸鲜酸辣
bright red color; smooth and tender paddy eel and vermicelli; salty, sour and spicy taste

鳝段粉丝
Paddy Eels with Pea Vermicelli

[原　料]
鳝鱼片150克　水发粉丝300克　食用油100克

[调料A]
泡辣椒末20克　郫县豆瓣20克　姜米10克　蒜米10克　葱花15克

[调料B]
食盐2克　胡椒粉1克　酱油6克　料酒15克　鸡精2克　鲜汤500克

[调料C]
醋20克

[制　作]

1. 鳝鱼片洗净，切成长约10厘米的段，入沸水中煮30秒钟捞出。
2. 锅中放食用油烧至120℃，下调料A炒香，加调料B烧沸出味，沥去料渣，放入鳝段和粉丝烧入味，加调料C后起锅装碗，撒上少许葱花即成。

Ingredients
150g paddy eels chunks, 300g water-soaked Chinese pea vermicelli, 100g cooking oil

Seasonings A
20g pickled chilies (minced), 20g chili bean paste, 10g ginger (finely chopped), 10g garlic (finely chopped), 15g scallion (finely chopped)

Seasonings B
2g salt, 1g ground white pepper, 6g soy sauce, 15g Shaoxing cooking wine, 2g chicken essence granules, 500g everyday stock

Seasonings C
20g vinegar

Preparation

1. Rinse the paddy eel chunks, chop into 10cm-long sections and blanch for 30 seconds in boiling water.
2. Heat oil in a wok to 120℃, add Seasonings A and stir-fry till aromatic. Add Seasonings B, bring to a boil and continue to boil to bring out the fragrance. Remove the scums. Add the paddy eel sections and vermicelli, then braise so that the paddy eels and vermicelli absorb the flavors of the condiments. Season with Seasonings C, remove from the heat and transfer to a serving bowl. Sprinkle with chopped scallion.

肉质干香细嫩，麻辣鲜香味厚

tender and aromatic paddy eel; pungent and spicy taste

干煸鳝丝
Dry-Fried Paddy Eel Slivers

[原 料]
鳝鱼片300克 芹黄100克 食用油75克

[调 料]
郫县豆瓣25克 姜丝5克 蒜丝6克 蒜苗丝10克 食盐1克 酱油5克 醋2克 料酒15克 味精2克 花椒粉2克 芝麻油5克

[制 作]
1. 鳝鱼片洗净，切成长约8厘米、粗约0.4厘米的丝；芹黄切成长约5厘米的节。
2. 锅置火上，入食用油烧至180℃，下鳝丝、食盐、醋煸炒至棕色，加郫县豆瓣、料酒、姜丝、蒜丝炒出香味，放入酱油、芹黄炒断生，再下味精、芝麻油、蒜苗丝炒匀，起锅装盘，撒上花椒粉即成。

Ingredients
300g paddy eel slices, 100g tender celery, 75g cooking oil

Seasonings
25g Pixian chili bean paste, 5g ginger (shredded), 6g garlic (shredded), 10g baby leeks (shredded), 1g salt, 5g soy sauce, 2g vinegar, 15g Shaoxing cooking wine, 2g MSG, 2g ground roasted Sichuan pepper, 5g sesame oil

Preparation
1. Wash the paddy eel thoroughly, and cut into slivers about 8cm long and 0.4cm thick. Chop the celery into 5cm-long sections.
2. Heat oil in a wok to 180℃, add the paddy eel strips, salt and vinegar, and then stir-fry till browned. Add the chili bean paste, Shaoxing cooking wine, ginger and garlic, and stir-fry to bring out the aroma. Blend in the soy sauce and celery, and stir-fry till al dente. Season with MSG, sesame oil and leek. Mix well, remove from the stove, transfer to a serving dish and sprinkle with ground roasted Sichuan pepper.

大蒜烧鳝鱼

Braised Paddy Eels with Garlic

色泽红亮，质地嫩滑，鲜香微辣

bright color; tender paddy eel; aromatic and subtly hot taste

[原　料]
鳝鱼片200克　大蒜100克　食用油80克

[调　料]
郫县豆瓣20克　姜片3克　葱段20克　食盐1克　味精2克　酱油5克　料酒15克　水淀粉15克

[制　作]
1. 鳝鱼片洗净，切成长约6厘米的段。
2. 炒锅置旺火，下食用油烧至180℃，下鳝鱼段煸炒至卷缩、表面水分干，续下郫县豆瓣、姜片、大蒜、葱段炒香，再下鲜汤、食盐、酱油、料酒烧至蒜软，最后下味精、水淀粉，待汁浓亮油时起锅装盘即成。

Ingredients
200g paddy eels, 100g garlic, 80g cooking oil

Seasonings
20g Pixian chili bean paste, 3g ginger (sliced), 20g scallion (cut into sections), 1g salt, 2g MSG, 5g soy sauce, 15g Shaoxing cooking wine, 15g cornstarch-water mixture

Preparation
1. Wash the paddy eels. Cut into 6cm lengths.
2. Heat oil in a wok to 180℃, add paddy eels, and stir-fry till the paddy eel pieces curl up and the skin dries. Add Pixian chili bean paste, ginger, garlic and scallion, and stir-fry till aromatic. Add the stock, salt, soy sauce and Shaoxing cooking wine, braise till the garlic is soft, add the MSG and cornstarch-water mixture. Wait till the sauce is thick and lustrous. Remove from the flame and transfer to a serving dish.

Hot-and-Spicy Loach
香辣炮泥鳅

肉质细嫩，咸鲜微辣
tender loach; salty, savoury and slightly hot taste

[原 料]
大泥鳅400克

[调 料]
大蒜30克 泡辣椒20克 干辣椒段20克 花椒10克 姜片10克 葱段100克 郫县豆瓣15克 食盐2克 味精2克 料酒15克 食用油200克 芝麻油5克 鲜汤500克

[制 作]
1. 泥鳅去内脏后洗净；泡辣椒剁细。
2. 锅置火上，入食用油烧至120℃，放郫县豆瓣、泡辣椒、干辣椒段、花椒、姜片、葱段炒出香味，加鲜汤烧沸，放入泥鳅、料酒、大蒜，用小火焖煮至泥鳅酥软，再下味精、芝麻油后起锅即成。

Ingredients
400g muddy loaches

Seasonings
30g garlic, 20g pickled chilies, 20g dried chilies (cut into sections), 10g Sichuan pepper, 10g ginger (sliced), 100g scallion (cut into sections), 15g Pixian chili bean paste, 2g salt, 2g MSG, 15g Shaoxing cooking wine, 200g cooking oil, 5g sesame oil, 500g everyday stock

Preparation
1. Core and rinse the loaches. Finely chop the pickled chilies.
2. Heat oil in a wok to 120℃, add Pixian chili bean paste, pickled chilies, dried chilies, Sichuan pepper, ginger, scallion and stir-fry until aromatic. Pour in some stock and bring to a boil. Add the loaches, Shaoxing cooking wine, garlic and simmer till the loaches become soft and tender. Season with MSG and sesame oil, and transfer to a serving dish.

典故

此菜特为清光绪年间四川总督丁宝桢而做,因丁宝桢曾官封太子少保,俗称"宫保",此菜据此而得名。

Note

The dish was originally made for Ding Baozhen, a governor of Sichuan during the Qing Dynasty, whose official title was "Gongbao".

[原 料]

鸡肉200克 酥花仁40克 食用油60克

[调料A]

食盐0.5克 料酒5克 酱油3克 水淀粉10克

[调料B]

食盐1克 料酒5克 酱油7克 醋10克 白糖10克 味精1克 水淀粉15克 鲜汤20克

[调料C]

干辣椒节10克 花椒4克 姜片8克 蒜片10克 葱丁15克

[制 作]

1. 鸡肉斩成约1.5厘米见方的丁,加入调料A拌匀。
2. 将调料B调匀成芡汁。
3. 锅中下食用油烧至140℃,下干辣椒、花椒炒香,续下鸡丁炒至断生,再下姜片、蒜片、葱丁炒香,倒入芡汁,待收汁亮油时放入酥花仁炒匀装盘即成。

Ingredients

200g chicken, 40g crispy peanuts, 60g cooking oil,

Seasonings A

0.5g salt, 5g Shaoxing cooking wine, 3g soy sauce, 10g cornstarch-water mixture,

Seasonings B

1g salt, 5g Shaoxing cooking wine, 7g soy sauce, 10g vinegar, 10g sugar, 1g MSG, 15g cornstarch-water mixture, 20g everyday stock,

Seasonings C

10g dried chilies (cut into sections), 4g Sichuan pepper, 8g ginger (sliced), 10g garlic (sliced), 15g scallion (finely chopped)

Preparation

1. Cut the chicken into $(1.5cm)^3$ cubes, add Seasonings A and mix well.
2. Mix Seasonings B to make the thickening sauce.
3. Heat oil in a wok to 140℃, add the dried chilies and Sichuan pepper, and stir-fry to bring out the aroma. Add the diced chicken, stir-fry till just cooked, and then add ginger, garlic, scallion and the thickening sauce. Add the cripsy peanuts when the sauce is thick and lustrous, mix evenly and then transfer to a serving dish.

色泽棕红，鸡丁细嫩，花仁酥脆，咸鲜甜酸，有干辣椒和花椒的香辣和香麻味

brownish and reddish color; tender chicken; crispy peanuts; a rich medley of sour, sweet, salty and zingy taste pepped up with Sichuan pepper and chili

宫保鸡丁
Gongbao Diced Chicken

太白鸡
Taibai Chicken

色泽金红发亮，肉质软糯，味咸鲜微辣

tender, juicy and glutinous chicken; lustrous and reddish color; salty, spicy and appetizing taste

典故

相传此菜是唐代大诗人李白在四川时酷爱吃的菜肴，因李白字"太白"而得名。

Note

It is said that this dish was the favorite of the famous poet Li Taibai when he was in Sichuan. That's why the dish was named after him.

[原　料]

鸡腿肉1000克　食用油1000克（约耗100克）

[调　料]

干辣椒段20克　泡辣椒段100克　八角1个　姜片15克　葱段20克　食盐5克　醪糟汁20克　白糖3克　料酒10克　胡椒粉1克　糖色20克　味精2克　芝麻油5克　鲜汤200克

[制　作]

1. 鸡肉去骨后斩成约4厘米见方的块，放入180℃的食用油中炸至表皮呈浅黄色时捞出。
2. 锅中下食用油烧至120℃，放入干辣椒、泡辣椒、姜片、葱段炒香，掺鲜汤烧沸，加入鸡块、八角、食盐、料酒、胡椒粉、白糖、糖色、醪糟汁烧沸，去掉浮沫，用小火烧至肉软熟、汁稠时捞出部分干辣椒、泡辣椒，再加入味精、芝麻油，收汁亮油后装盘即成。

Ingredients

1000g chicken leg meat, 1000g cooking oil for deep-frying

Seasonings

20g dried chilies, 100g pickled chilies (cut into sections), 1 star anise, 15g ginger (sliced), 20g scallion (cut into sections), 5g salt, 20g fermented glutinous rice wine, 3g sugar, 10g Shaoxing cooking wine, 1g ground white powder, 20g caramel color, 2g MSG, 5g sesame oil, 200g everyday stock.

Preparation

1. De-bone the chicken, cut into (4cm)³ cubes and deep-fry in 180℃ oil till brownish.
2. Heat oil in a wok to 120℃, add dried chilies, pickled chilies, ginger and scallion, and stir-fry to bring out the aroma. Add the stock, bring to a boil, add chicken, star anise, salt, Shaoxing cooking wine, ground white pepper, sugar, caramel color and fermented glutinous rice wine, bring to a second boil and remove scums. Simmer over a low flame till the chicken is soft and cooked through and the sauce beomes thick. Remove and discard some dried chilies and pickled chilies. Add the MSG and sesame oil. Wait till the sauce becomes thick and lustrous, and transfer to a serving dish.

色彩美观,鸡肉细嫩,玉米窝窝头松泡
beautiful and appealing color; tender chicken; puffy corn buns

Chopped Chicken with Steamed Corn Buns
鸡米杂粮配窝窝头

[原 料]

玉米窝窝头10个 鸡脯肉100克 罐头玉米100克 青、红尖椒粒各50克 食用油75克

[调 料]

食盐1克 甜酱15克 芝麻油5克 味精1克 水淀粉10克

[制 作]

1. 鸡脯肉切成粒,加食盐、水淀粉拌匀。
2. 锅置火上,下食用油烧至120℃,放入鸡肉粒炒散籽,续下甜酱炒匀,再下玉米粒、青、红尖椒炒断生,最后下味精、芝麻油和匀,起锅装盘后配上玉米窝窝头即成。

Ingredients

10 steamed corn buns, 100g chicken breast meat, 100g canned corn, 50g red chili peppers (finely chopped), 50g green chili peppers (finely chopped), 75g cooking oil

Seasonings

1g salt, 15g fermented flour paste, 5g sesame oil, 1g MSG, 10g cornstarch-water mixture.

Preparation

1. Finely chop the chicken breast, add salt and cornstarch-water mixture, and blend well.
2. Heat oil in a wok to 120℃, add the chicken and stir-fry till they no longer stick together. Add the fermented flour paste and stir well. Blend in the canned corn, green chili peppers and red chili peppers, and stir-fry till al dente. Add the MSG and sesame oil, blend well, remove from the stove and transfer to a serving dish. Serve with the corn buns.

Stewed Silkie Chicken with Wild Mushrooms
野生菌煨乌鸡

质地熟软,鲜美清爽
tender and soft chicken;
delicate and fragrant soup

[原　料]

乌鸡肉1000克　牛肝菌25克　鸡腿菇25克　香菇25克

[调　料]

食盐3克　料酒15克　姜片15克　葱段25克

[制　作]

1. 乌鸡肉斩成块,余水后备用;牛肝菌、鸡腿菇、香菇用清水浸泡后洗净。
2. 鸡块放入瓷罐中,下料酒、葱段、姜片、清水,加盖用旺火烧沸后改用小火煨制3小时,再放入牛肝菌、鸡腿菇、香菇、食盐煨制1小时即成。

Ingredients
1000g silkie chicken, 25g boletus, 25g shaggy mane, 25g shiitaki

Seasonings
3g salt, 15g Shaoxing cooking wine, 15g ginger (sliced), 25g scallion (cut into sections)

Preparation
1. Chunk the chicken and blanch. Soak the mushrooms and wash thoroughly.
2. Put the chunked chicken into an earthen pot, and add the Shaoxing cooking wine, scallion, ginger and water. Cover the pot and bring to a boil over a high flame. Simmer for 3 hours. Add the mushrooms and salt, and then simmer for another hour.

质地熟软，鲜美清爽
tender, soft and delicate chicken and mushrooms; delectable and savoury taste

松茸炖土鸡

Stewed Free-range Chicken with Matsutake

[原　料]

土鸡1500克　松茸50克

[调　料]

食盐5克　料酒15克　姜片15克　葱段25克

[制　作]

1. 土鸡宰杀后治净，斩成小块，入沸水锅中氽水后捞出洗净；松茸用清水浸泡后洗净。
2. 取大碗一个，放入鸡块，掺入清水，放料酒、葱段、姜片，取一张白纸淋湿后将碗盖上，入锅隔水炖制约2小时，最后放入松茸，调入食盐续炖5分钟即成。

Ingredients

1500g free-range chicken, 50g matsutake mushrooms,

Seasonings

5g salt, 15g Shaoxing cooking wine, 15g ginger (sliced), 25g scallion (cut into sections)

Preparation

1. Kill the chicken and wash thoroughly, chop into small cubes, blanch, remove and rinse. Soak the matsutake mushrooms in water for a while, and then rinse.
2. Get a big bowl, put the chicken cubes in, add water, Shaoxing cooking wine, scallion and ginger, and then cover the bowl with a wet piece of paper. Transfer the bowl to the water in a pot and stew for about 2 hours. Add the mushrooms and salt to the bowl and simmer for another 5 minutes.

芙蓉鸡片
Hibiscus-like Chicken

色白自然，质地柔软细嫩，味咸鲜

pure white color; tender and soft chicken; salty and delicate taste

[原 料]

鸡脯肉100克 冬笋片20克 熟火腿片10克 豌豆苗10克 食用油100克

[调 料]

鸡蛋清60克 食盐3克 味精1克 胡椒粉0.5克 料酒5克 水淀粉70克 鸡油5克 鲜汤100克

[制 作]

1. 将鸡脯肉捶成泥蓉状放入碗中，加清水、鸡蛋清、食盐、味精、胡椒粉、料酒、水淀粉调匀成鸡浆。
2. 炒锅置火上，入食用油烧至80℃，舀入鸡浆摊成片状，待熟时铲入鲜汤中浸泡以褪去油脂。
3. 锅内放鲜汤、鸡浆片、冬笋片、火腿片，加食盐、味精、胡椒粉烧沸出味，放入豌豆苗，用水淀粉勾成清二流芡，淋鸡油推匀，起锅装盘即成。

Ingredients

100g chicken breast meat, 20g winter bamboo shoots (sliced), 10g pre-cooked ham (sliced), 10g pea vine sprouts, 100g cooking oil

Seasonings

60g egg white, 3g salt, 1g MSG, 0.5g ground white pepper 5g Shaoxing cooking wine, 70g cornstarch-water mixture, 5g chicken oil, 100g everyday stock,

Preparation

1. Mince the chicken, transfer to a bowl, and add water, egg white, salt, MSG, pepper, Shaoxing cooking wine and cornstarch-water mixture to make a chicken paste.
2. Heat oil in a wok to 80℃, and add small spoonfuls of chicken paste, flattening them gently. Wait till the flattened chicken pieces are fully cooked, and then transfer into the stock to remove any grease.
3. Mix the stock, chicken pieces, winter bamboo shoots, ham, salt, MSG and ground white pepper in a wok, bring to a boil, add the pea vine sprouts, cornstarch-water mixture and chicken oil, stir well and transfer to a serving dish.

色泽棕红，鸡肉鲜嫩，辣味突出
red-brown color; tender chicken; pungent and spicy flavors

辣子鸡丁
Diced Chicken with Pickled Chilies

[原 料]
鸡腿肉150克 青笋100克 食用油75克

[调 料]
泡辣椒末25克 姜片3克 蒜片5克 葱丁10克 食盐2克 料酒5克 酱油5克 醋5克 白糖3克 味精1克 鲜汤30克 水淀粉25克

[制 作]
1. 鸡肉切成2厘米见方的丁，加食盐、料酒、水淀粉拌匀；青笋切成约1.5厘米见方的丁；将食盐、酱油、醋、白糖、味精、鲜汤、水淀粉调匀成芡汁。
2. 锅置旺火上，下食用油烧热，放入鸡丁炒至断生，续下泡辣椒末、姜片、蒜片、葱丁炒香，最后加入青笋丁，倒入芡汁，待收汁亮油后起锅即成。

Ingredient
200g chicken leg meat, 100g asparagus lettuce, 75g cooking oil,

Seasonings
25g pickled chilies (minced), 3g ginger (sliced), 5g garlic (sliced), 10g scallion (finely chopped), 2g salt, 5g Shaoxing cooking wine, 5g soy sauce, 5g vinegar, 3g sugar, 1g MSG, 30g everyday stock, 25g cornstarch-water mixture

Preparation
1. Cut the chicken into (2cm)^3 cubes, add the salt, Shaoxing cooking wine and cornstarch-water mixture, and mix well. Cut the asparagus lettuce into 1.5cm^3 cubes. Mix the salt, soy sauce, vinegar, sugar, MSG, stock and cornstarch-water mixture to make thickening sauce.
2. Heat a wok over a high flame, add the oil and wait till it is hot. Add the chicken cubes and stir-fry till just cooked. Add the pickled chili, ginger, garlic and scallion, and continue to stir-fry to bring out the flavors. Blend in the asparagus lettuce and the thickening sauce. Wait till the sauce becomes thick and lustrous, remove from the heat and transfer to a serving dish.

鸡豆花 Chicken Curd

汤清菜白，清鲜味醇，质地细嫩
clear soup; savoury taste; tender and mild chicken

鸡豆花
Chicken Curd

[原 料]
鸡脯肉400克 菜心20克 鸡蛋清300克 清汤2500克

[调 料]
食盐2克 水淀粉50克

[制 作]
1. 鸡脯肉切成小条，放入搅打器内加少许清水搅拌成泥蓉状，倒入大碗中，加清水、鸡蛋清、食盐、水淀粉搅匀成鸡浆。菜心入沸水中氽熟备用。
2. 锅置火上，掺入清汤烧沸，倒入鸡浆，用小火慢煮至熟，舀入汤碗，灌入清汤，放上氽熟的菜心即成。

Ingredients
400g chicken breast meat, 20g tender leafy vegetables, 300g egg white, 2500g consomme

Seasonings
2g salt, 50g cornstarch-water mixture

Preparation
1. Cut the chicken into strips and transfer to a mixer. Add water into the mixer and stir into paste. Transfer the paste into a big bowl, add water, egg white, salt and cornstarch-water mixture, and then stir into chicken batter. Blanch the vegetables and lay aside.
2. Put a wok over a flame, add the consommé and bring to a boil. Pour in the chicken batter, simmer till fully cooked and then ladle into a soup bowl. Pour some consomme into the soup bowl and put the blanched vegetables onto the chicken.

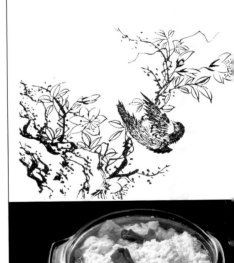

白绿相映，质地细嫩，汤汁清澈，咸鲜清淡
contrasting colors of white and verdant; tender texture; clear and delicate soup; mild and yummy taste

鸡蒙葵菜

Cluster Mallow Coated with Chicken Mince

[原 料]

鸡脯肉200克 葵菜心100克 鸡蛋清150克 清汤1500克

[调 料]

食盐1克 水淀粉35克 清水100克 化猪油100克

[制 作]

1. 鸡脯肉切成小条，放入搅打器内加鸡蛋清、食盐、清水、化猪油、水淀粉搅打成色白、松泡、细腻的鸡糁。
2. 葵菜心入沸水中氽一水后捞出沥干水分，用鸡糁将葵菜心包裹成型，放入热的清汤中；待所有葵菜心都包裹完后，与清汤一起倒入锅中，用小火加热煮熟，捞出后放入汤碗中，灌入清汤即成。

Ingredients

200g chicken breast meat, 100g tender cluster mallow, 150g egg white, 1500g consomme

Seasonings

1g salt, 35g cornstarch-water mixture, 100g lard

Preparation

1. Cut the chicken breast into strips, and transfer into a mixer. Add the egg white, salt, water, lard and cornstarch-water mixture, and then stir into white, puffy and tender chicken paste.
2. Blanch the cluster mallow in water, remove and drain. Wrap every leaf of the cluster mallow with chicken paste and transfer into hot consomme. Pour the cluster mallow into a wok together with the consomme, simmer over a low flame till fully cooked. Remove the cluster mallow from the wok, transfer to a soup bowl, and pour the consommé in the wok into the bowl.

白果炖鸡

Stewed Chicken with Gingko Nuts

鸡肉细嫩醇鲜，白果软糯回甜，味醇、味厚、味美

tender chicken; soft and slightly glutinous gingko nuts; delicate and savory soup

典故

此菜为青城山道家所创制，被誉为青城山"四绝"之一。

Note

Invented by the Taoists of the Qingcheng Mountain, this dish is renowned as one of "the Four Unrivaled of the Qingcheng Mountain".

[原 料]

净仔母鸡1只（约1000克）　白果150克　鲜汤1250克

[调 料]

生姜25克　大葱80克　食盐5克　胡椒粉1.5克　味精2克　料酒20克

[制 作]

1．白果去壳、皮和芯，用清水浸泡待用。
2．净仔母鸡入沸水锅中汆水后捞出洗净，放入砂锅中，加生姜、大葱、料酒、鲜汤，用微火炖制2~3小时，再加入白果、食盐、胡椒粉、味精，继续炖约半小时即成。

Ingredients

1 poussin hen (about 1000g), 150g fresh gingko nuts, 1250g everyday stock

Seasonings

25g ginger, 80g scallion, 5g salt, 1.5g ground white pepper 2g MSG, 20g Shaoxing cooking wine

Preparation

1. Shell, peel and core the gingko nuts, and then soak in water.
2. Blanch the poussin hen, remove, rinse and transfer to an earthen pot. Add ginger, scallion, Shaoxing cooking wine and stock to the pot and then simmer for 2 to 3 hours. Add the gingko nuts, salt, pepper and MSG and continue to stew for another 30 minutes.

色泽棕黄，鸡肉熟软，咸鲜清香

golden brown color; soft and glutinous chicken; aromatic and delicate taste

黄焖鸡
Golden Chicken Stew

[原 料]

仔公鸡肉300克 青笋150克 食用油1000克（约耗70克）

[调 料]

食盐2.5克 料酒20克 姜8克 葱15克 味精1克 鲜汤500克 水淀粉15克 糖色20克

[制 作]

1. 将鸡斩成约4厘米见方的块；姜拍破、葱绾结。青笋切成滚刀块后氽水断生，漂凉保色待用。
2. 锅中下食用油烧至180℃，放入鸡块炸1分钟后捞出，入铝锅内加鲜汤烧沸，撇去浮沫，下食盐、糖色、料酒、姜、葱，加盖后转中小火焖煮至熟，续下青笋焖至熟软，捡去姜、葱，放味精，用水淀粉勾成二流芡，起锅装盘即成。

Ingredients

300g poussin rooster meat, 150g asparagus lettuce, 1000g cooking oil for deep-frying

Seasonings

2.5g salt, 20g Shaoxing cooking wine, 8g ginger, 15g scallion, 1g MSG, 500g everyday stock, 15g cornstarch-water mixture, 20g caramel color

Preparation

1. Cut the asparagus lettuce into chunks and the chicken into 4cm³ cubes. Smash the ginger with the flat side of the blade, and tie the scallion into knots. Blanch the asparagus lettuce in water till al dente, and rinse in cold water till cool so as to preserve its natural color.
2. Heat oil in a wok to 180℃, add chicken cubes and deep-fry for 1 minute. Remove the chicken, transfer to an aluminum pot, add the stock and bring to a boil. Remove the scums, add the salt, caramel color, Shaoxing cooking wine, ginger and scallion, cover and simmer over a medium-low flame till the chiken is fully cooked. Add the asparagus lettuce, cover and simmer till soft. Remove and discard the ginger and scallion, add MSG and cornstarch-water mixture, remove from the stove and transfer to a serving dish.

外酥香，内细嫩，咸甜酸辣兼备，姜葱蒜味浓郁
tender chicken that are crispy on the outside; sweet, sour and sligtly hot taste; strong aroma of garlic, scallion and ginger

鱼香八块鸡
Chicken Chunks in Fish-Flavor Sauce

[原 料]
鸡肉200克 食用油1000克（约耗100克）

[调 料]
泡辣椒末30克 姜末6克 葱花15克 蒜米10克 食盐4克 料酒10克 白糖20克 酱油5克 醋15克 味精3克 全蛋淀粉120克 水淀粉20克 鲜汤50克

[制 作]
1. 鸡肉斩成2.5厘米见方的块，加食盐、料酒、全蛋淀粉拌匀；食盐、料酒、白糖、酱油、醋、味精、水淀粉、鲜汤调成芡汁。
2. 锅中放入食用油烧至170℃，放入鸡块炸至表面金黄时捞出。
3. 锅中留油，加入泡辣椒末、姜末、葱花、蒜米炒香，倒入芡汁收稠汤汁，放入鸡块炒匀后起锅装盘即可。

Ingredients
200g chicken, 1000g cooking oil for deep-frying,

Seasonings
30g pickled chilies (finely chopped), 6g ginger (finely chopped), 15g scallion (finely chopped), 10g garlic (finely chopped), 4g salt, 10g Shaoxing cooking wine, 20g sugar, 5g soy sauce, 15g vinegar, 3g MSG, 120g mixture of egg and cornstarch, 20g cornstarch-water mixture, 50g everyday stock,

Preparation
1. Cut the chicken into 2.5cm³ cubes, add some salt, Shaoxing cooking wine, and mixture of egg and cornstarch, then blend well. Mix salt, Shaoxing cooking wine, sugar, soy sauce, vinegar, MSG, cornstarch-water mixture and stock to make thickening sauce.
2. Heat oil in a wok to 170°C and deep-fry the chicken cubes till their surfaces become golden brown.
3. Heat oil in a wok, add pickled chilies, ginger, scallion and garlic, and stir-fry till aromtic. Pour in the thickening sauce and chicken cubes, blend well and transfer to a serving dish.

Chicken Chunks in Fish-Flavor Sauce

鱼香八块鸡

雪花鸡淖
Snowy Chicken

色泽洁白如雪，质地细嫩滑爽，入口咸鲜清淡

pure snowy color; tender and smooth chicken; salty and delicate taste

雪花鸡淖
Snowy Chicken

[原　料]

鸡脯肉150克　熟火腿5克　鸡蛋清100克　食用油125克

[调　料]

水淀粉75克　食盐3克　姜葱水100克　味精2克　热鲜汤150克

[制　作]

1. 鸡蛋清搅打成泡状；熟火腿切成末。
2. 鸡肉去皮、筋，捣蓉后装入碗内，分次加入姜葱水澥散，再加入鸡蛋清、食盐、味精、水淀粉搅成糊状。
3. 锅置旺火上，下食用油烧至150℃，在鸡糊内加入热鲜汤调匀，倒入锅内，翻炒松散至色白质嫩时起锅，装入盘内，撒上火腿末即成。

Ingredients

150g chicken breast meat, 5g precooked ham, 100g egg white, 125g cooking oil,

Seasonings

75g cornstarch-water mixture, 3g salt, 100g ginger-and-scallion-flavored water, 2g MSG, 150g hot everyday stock

Preparation

1. Beat the egg white to product foams. Finely chop the ham.
2. Remove the skin and tendon of the chicken, mince and transfer to a bowl. Add the ginger-and-scallion-flavord water in successive portions and stir well. Add the egg white, salt, MSG and cornstarch-water mixture to make paste.
3. Heat oil in a wok to 150℃, mix chicken paste with the hot stock, and pour into the wok. Keep stiring till the chicken whitens and doesn't stick together. Transfer to a serving dish and sprinkle with chopped ham.

色彩鲜艳，质地脆嫩，香辣可口
contrasting colors; tender and crunchy meat and vegetables; spicy and appetizing taste

香辣掌中宝
Hot-and-Spicy Chicken Feet Pad

[原 料]
掌中宝（鸡脚掌肉）200克 青、红辣椒节100克 食用油500克（约耗30克）

[调料A]
食盐1克 酱油6克

[调料B]
干辣椒段2克 姜片5克 蒜片5克 葱丁15克 香辣酱25克

[调料C]
花椒油3克 蚝油10克 味精1克 白糖2克

[制 作]
1．掌中宝与调料A拌匀、码味，入150℃的食用油中炸熟后捞出备用。
2．锅内留油少许，放入调料B炒香，再放入掌中宝、辣椒节和调料C翻炒成熟，起锅装盘即成。

Ingredients
200g pad meat of chicken feet, 100g red chili peppers (cut into sections), 100g green chili peppers (cut into sections), 500g cooking oil

Seasonings A
1g salt, 6g soy sauce

Seasonings B
2g dried chilies (segmented), 5g ginger (sliced), 5g garlic (sliced), 15g scallion (finely chopped), 25g chili pepper paste

Seasonings C
3g Sichuan pepper oil, 10g oyster sauce, 1g MSG, 2g sugar

Preparation
1. Mix the pad meat of chicken feet with Seasonings A, blend well and marinate. Fry in 150℃ oil till cooked through, and then remove.
2. Leave a dash of oil in the wok, add Seasonings B and stir-fry till fragrant. Add the pad meat, red chili peppers, green chili peppers and Seasonings C, and stir-fry till cooked through. Transfer to a serving dish.

青椒鸡杂

Chicken Hotchpotch with Green Peppers

鸡杂脆爽，咸鲜麻辣
crunchy and savory chicken offal; lingering aroma

青椒鸡杂

Chicken Hotchpotch with Green Peppers

[原 料]
鸡杂400克 青尖椒粒150克 食用油80克

[调 料]
泡辣椒段60克 香辣酱30克 鲜青花椒40克 姜片10克 蒜片10克 葱段15克 食盐3克 芝麻油5克 料酒15克 味精2克 胡椒粉1克

[制 作]
1. 鸡杂清洗干净后切片，入沸水中余尽血水后捞出备用。
2. 锅中下食用油烧至150℃，放青尖椒粒、泡辣椒段、鲜青花椒、香辣酱、姜片、蒜片、葱段炒香，再放入鸡杂、食盐、料酒、味精、胡椒粉、芝麻油炒匀，起锅装盘即成。

Ingredients
400g chicken offal, 150g green chili peppers (chopped), 80g cooking oil,

Seasonings
60g pickled chilies (cut into sections), 30g chili pepper paste, 40g green Sichuan pepper, 10g ginger (sliced), 10g garlic (sliced), 15g scallioin (cut into sections), 3g salt, 5g sesame oil, 15g Shaoxing cooking wine, 2g MSG, 1g ground white pepper

Preparation
1. Wash thoroughly the chicken offal, and cut into slices. Blanch in water to remove any remaining blood, and remove.
2. Heat oil in a wok to 150℃, add green chili peppers, pickled chilies, green Sichuan pepper, chili pepper paste, ginger, garlic and scallion, and stir-fry to bring out the fragrance. Add the offal, salt, Shaoxing cooking wine, MSG, ground white pepper and sesame oil, stir well and transfer to a serving dish.

汤清味醇，软糯适口
clear and savoury soup; soft and glutinous duck

虫草鸭子
Steamed Duck with Caterpillar Fungus

[原 料]
公鸭1只 虫草10克

[调 料]
姜片10克 葱段15克 食盐2克 胡椒粉1克 料酒20克 鸡精2克

[制 作]
1. 虫草用清水浸泡后用刷子刷洗干净。
2. 公鸭初加工后去掉脚爪，入锅煮尽血水后捞出，在鸭腹两边用小刀各戳5个小口，将虫草插入鸭腹内。
3. 将鸭腹朝上放在大汤碗内，掺入清水，放入食盐、胡椒粉、料酒、姜片、葱段，用湿绵纸封严碗口，入笼蒸约2小时至鸭肉熟烂，拣去姜片、葱段，加入鸡精即成。

Ingredients
one drake, 10g caterpillar fungus,

Seasonings
10g ginger (sliced), 15g scallion (cut into sections), 2g salt, 1g ground white pepper, 20g Shaoxing cooking wine, 2g chicken essence granules

Preparation
1. Soak the caterpillar fungus in water for a while and brush to clean thoroughly.
2. Remove the duck's feet, and blanch to clean away any remaining blood. Make five cuts into each side of the duck's belly and insert a piece of the caterpillar fungus into each cut.
3. Lay the duck on its back in a large soup bowl, add the water, salt, ground white pepper, Shaoxing cooking wine, ginger, and scallion, and then seal by covering the bowl with wet cotton paper towels. Transfer the bowl to a steamer and steam for about 2 hours till the duck is cooked through. Remove and discard the ginger and scallion, and add chicken essence granules.

樟茶鸭
Tea-Smoked Duck

皮酥肉嫩，香醇味长，滋润不腻
tender meat wrapped in crispy skin; lasting fragrance; aromatic but not greasy

樟茶鸭
Tea-Smoked Duck

[原 料]
公鸭1只 食用油1500克（约耗50克）

[调 料]
食盐4克 花椒5克 料酒20克 胡椒粉1克 醪糟汁20克 芝麻油5克

[制 作]
1. 将调料混合后均匀地抹在鸭身上，腌制8小时后用沸水汆水后捞出。
2. 揾干鸭身水分，用茉莉花茶、樟树叶、稻草、松柏枝作熏料，熏至鸭皮呈黄色时出炉，装入大碗，入笼蒸2小时后取出晾凉。
3. 锅中下食用油烧至200℃，放入鸭子炸至表皮酥香时取出，斩成块后装盘即成。

Ingredients
A drake, 1500g cooking oil for deep-frying

Seasonings
4g salt, 5g Sichuan pepper, 20g Shaoxing cooking wine, 1g ground white pepper, 20g fermented glutinous rice wine, 5g sesame oil

Preparation
1. Mix the Seasonings, coat the drake with the mixture, marinate for 8 hours, and then blanch.
2. Drain the duck, and then smoke the duck with fuels consisting of jasmine tea, camphor laurel leaves, straws, pine and cypress branches until the skin of the duck becomes brownish. Transfer the the duck to a large bowl, steam for two hours, remove and cool.
3. Heat the oil in a wok to 200℃, and deep-fry the duck until its skin is aromatic and crispy. Remove the duck from the oil, chop into chunks and transfer to a serving dish.

甜皮鸭
Crispy Sweet-Skinned Duck

色泽棕红光亮，肉质细嫩，香味浓郁

bright brown color; tender meat; lingering aroma

[原 料]
土鸭1只 食用油1500克（约耗100克）

[调 料]
姜片20克 葱段20克 食盐30克 饴糖40克 糖色40克 料酒40克 花椒5克 香料30克

[制 作]
1. 土鸭初加工后用花椒、食盐、料酒抹遍鸭身内外，腌制5小时；香料用纱布包成香料包。
2. 锅内加入鲜汤、食盐、料酒、糖色、姜片、葱段、香料包，大火烧沸后改用小火熬成卤水，放入腌好的鸭子，用中小火煮至鸭肉软熟后捞出沥干水分。
3. 将鸭子放入热油锅中炸至皮酥、色呈棕红时捞出，刷上饴糖即成。

Ingredients
1 free-range duck, 1500g cooking oil for deep-frying, 2000g stock,

Seasonings
20g ginger (sliced), 20g scallion (cut into sections), 30g salt, 40g maltose, 40g caramel color, 40g Shaoxing cooking wine, 5g Sichuan pepper, 30g mixed herbal spices

Preparation
1. Smear the duck with Sichuan pepper, salt and the Shaoxing cooking wine, and marinate for 5 hours. Wrap up the mixed herbal spices in cheesecloth to make a seasoning packet.
2. Put the stock, salt, Shaoxing cooking wine, caramel color, ginger, scallion, the seasoning packet into a pot. Bring to a boil and simmer to make Sichuan-style broth. Put the marinated duck into the pot and stew over a medium-low flame till soft and cooked through. Remove and drain.
3. Deep-fry the duck till the skin becomes crispy and brown. Remove and coat with maltose.

色泽深红，味浓鲜香
dark brown color, aromatic and appetizing taste

姜爆鸭丝
Quick-Fried Duck Slivers with Ginger

[原　料]

樟茶鸭肉200克　仔姜75克　甜椒75克　蒜苗30克　食用油75克

[调　料]

酱油5克　白糖2克　味精1克　芝麻油5克

[制　作]

1. 鸭肉、甜椒切成长约6厘米的粗丝；仔姜切成长约4厘米的细丝；蒜苗切成长约4厘米的粗丝。
2. 锅置火上，下食用油烧至150℃，入鸭丝炒香，再放甜椒丝、仔姜丝炒出香味，续下蒜苗、酱油、白糖、味精炒至蒜苗断生，加芝麻油炒匀起锅即成。

Ingredients

200g duck smoked with jasmine tea and camphor laurel leaves, 75g tender ginger, 75g red bell peppers, 30g baby leeks, 75g cooking oil

Seasonings

5g soy sauce, 2g sugar, 1g MSG, 5g sesame oil

Preparation

1. Cut the duck and bell peppers into 6cm-long strips, tender ginger into 4cm-long slivers, and leeks into 4cm-long strips.
2. Heat oil in a wok to 150℃. Add the duck strips and stir-fry to bring out the aroma. Add the bell peppers and ginger. Stir-fry till fragrant. Blend in the leeks, soy sauce, sugar, MSG and stir-fry till the leeks are just cooked. Blend in the sesame oil and mix well. Transfer to serving dish.

酱爆鸭舌

Quick-Fried Duck Tongues with Fermented Flour Paste

色泽棕红，酱香浓郁
lustrous brown color; strong aroma of fermented flour paste

酱爆鸭舌

Quick-Fried Duck Tongues with Fermented Flour Paste

[原料]
卤鸭舌400克 菜心50克 食用油1000克（约耗45克）

[调料]
甜面酱20克 料酒10克 酱油5克 味精2克

[制作]
1. 菜心洗净，入沸水中氽熟后捞出装盘。
2. 锅置旺火上，下食用油烧至150℃，入卤鸭舌炸至色微黄时捞出；锅内留油少许烧至100℃，入甜面酱炒出香味，再放入鸭舌、料酒爆炒，最后下味精、酱油炒匀，出锅盛于菜心上即成。

Ingredients
400g duck tongues (pre-cooked in Sichuan-style broth), 50g tender leafy vegetables, 1000g oil for deep-frying

Seasonings
20g fermented flour paste, 10g Shaoxing cooking wine, 5g soy sauce, 2g MSG

Preparation
1. Wash the vegetables, blanch and transfer to a serving dish.
2. Heat the oil in a wok to 150℃, fry the duck tongues till slightly golden, and then remove. Leave a dash of oil in the wok, heat to 100℃, add the fermented flour paste and stir-fry till aromatic. Add the duck tongues and Shaoxing cooking wine, and stir-fry over a high flame. Add the MSG and soy sauce, stir well. Transfer the duck tongues to the serving dish with the vegetables at the bottom.

色泽红亮，外酥内嫩，麻辣鲜香
bright and lustrous color; aromatic and tender meat; spicy and pungent taste

香辣鸭唇
Hot-and-Spicy Duck Jaws

[原　料]
鸭唇750克　红甜椒50克　青椒50克　洋葱50克　食用油1000克（约耗70克）

[调　料]
卤水2000克　姜30克　葱段30克　料酒40克　辣椒粉20克　花椒粉4克　食盐2克　白糖3克　味精2克　芝麻油7克　花椒油3克

[制　作]
1. 鸭唇洗净，入沸水锅中余尽血水后捞出；红甜椒、青椒、洋葱均切成小颗粒。
2. 锅置火上，放入卤水、鸭唇、姜、葱段、料酒烧沸，改用小火卤20分钟至鸭唇软熟入味后捞出，晾干水分，放入180℃的食用油中炸至酥香后捞出。
3. 锅中下食用油烧至120℃，放入红甜椒、青椒、洋葱、辣椒粉、花椒粉、白糖、味精、芝麻油、花椒油炒香，加入鸭唇裹匀，起锅装盘即成。

Ingredients
750g duck lower jaws with tongues, 50g red bell peppers, 50g green chili peppers, 50g onions, 1000g cooking oil for deep-frying

Seasonings
2000g Sichuan-style broth, 30g ginger, 30g scallion (cut into sections), 40g Shaoxing cooking wine, 20g ground chilies, 4g ground roasted Sichuan pepper, 2g salt, 3g sugar, 2g MSG, 7g sesame oil, 3g Sichuan pepper oil,

Preparation
1. Rinse the duck jaws and blanch to remove any remaining blood. Chop the red bell peppers, green chili peppers and onions into grains.
2. Put a wok over a flame, and add the Sichuan-style broth, duck jaws, ginger, scallion and Shaoxing cooking wine. Bring to a boil and simmer for 20 minutes till the duck jaws are soft and cooked through. Remove, drain and deep-fry in 180℃ oil till crispy and aromatic.
3. Heat some oil in a wok to 120℃. Add the red bell peppers, green chili peppers, onions, ground chilies, ground roasted Sichuan pepper, sugar, MSG, sesame oil and Sichuan pepper oil, and stir-fry to bring out the fragrance. Blend in the duck jaws, stir well and transfer to a serving dish.

色泽金黄，香气浓郁，酥软适口
golden brown color; aromatic and lingering smell; crispy skin and soft meat

[原　料]
嫩土鸭1500克　食用油1500克（约耗80克）

[调　料]
姜片10克　葱段20克　食盐10克　花椒5克　五香粉3克　料酒35克　葱酱味碟1碟

[制　作]
1. 嫩土鸭初加工后用食盐、料酒、姜片、葱段、五香粉、花椒抹匀腌制2小时，放入蒸碗内上笼蒸至软熟，取出晾干水分。
2. 锅置旺火上，下食用油烧至210℃，放入鸭子炸至皮酥、色金黄时捞起，砍成条，摆入盘内，与葱酱味碟同时上桌即成。

Ingredients
1500g free-range duck, 1500g cooking oil for deep-frying

Seasonings
10g ginger (sliced), 20g scallion (cut into sections), 10g salt, 5g Sichuan pepper, 3g five-spice powder, 35g Shaoxing cooking wine, dipping sauce made of scallion and fermented flour paste (served on a saucer)

Preparation
1. Clean the duck. Smear the duck with salt, Shaoxing cooking wine, ginger, scallion, five-spice powder and Sichuan pepper, and leave to marinate for two hours. Lay the duck into a steaming bowl, transfer to a steamer, and steam till the duck is soft and cooked through. Remove and drain.
2. Heat oil in a wok to 210°C and deep-fry the duck till its skin becomes crispy and golden brown. Remove from the oil, cut into strips, and lay neatly onto a serving dish. Serve with the dipping sauce.

香酥鸭子
Crispy Duck

色泽淡雅，鲜嫩清淡
delicate in both color and taste

鸡㙡烩鸭腰

Braised Duck Kidneys with Collybia Mushrooms

[原　料]
鸭腰300克　鸡㙡菌200克　化猪油30克

[调　料]
姜片10克　葱段15克　食盐3克　胡椒粉1克　鲜汤250克　料酒10克　味精1克　水淀粉20克　化鸡油20克

[制　作]
1. 鸭腰煮熟，对剖撕皮后盛碗内；鸡㙡去粗皮、老筋，洗净菌脚，切薄片，入沸水中氽熟后备用。
2. 锅置火上，下化猪油烧至120℃，放入姜片、葱段炒香，掺入鲜汤烧沸，放入鸭腰、鸡㙡片、食盐、胡椒粉、料酒烩2分钟，入味精、水淀粉、化鸡油收汁浓稠，起锅装盘即成。

Ingredients
300g duck kidneys, 200g collybia mushrooms, 30g lard

Seasonings
10g ginger (sliced), 15g scallion (cut into sections), 3g salt, 1g ground white pepper, 10g Shaoxing cooking wine, 1g MSG, 20g cornstarch-water mixture, 20g chicken oil, 250g everyday stock

Preparation
1. Boil the duck kidneys till cooked through. Halve the kidneys, remove the membrane and transfer to a bowl. Remove the rough skin, strings and roots of the mushrooms. Rinse, slice and blanch the mushrooms till cooked through.
2. Heat lard in a wok to 120℃ and stir-fry ginger and scallion to bring out the aroma. Add the stock, bring to a boil and then remove and discard the ginger and scallion. Add duck kidneys, mushrooms, salt, ground white pepper and Shaoxing cooking wine, boil for 2 minutes and add the MSG, cornstarch-water mixture and chicken oil. Wait till the sauce thickens, remove from the stove and transfer to a serving dish.

天麻乳鸽

Pigeon Stew with Gastrodia Tuber

汤汁清澈，鸽肉细嫩，咸鲜醇香
clear soup; tender pigeon meat; soft and glutinous gastrodia tuber; delicate and aromatic taste

天麻乳鸽
Pigeon Stew with Gastrodia Tuber

［原　料］
乳鸽2只　天麻50克

［调　料］
食盐2克　鸡精2克　胡椒粉1克　姜片10克　葱白15克　料酒15克

［制　作］
1. 乳鸽初加工后入沸水中氽水备用；天麻加清水上笼蒸软，切成薄片备用。
2. 在大汤碗中注入清水，放入乳鸽、天麻、姜片、葱白、料酒、胡椒粉，加盖后放入笼中蒸至酥软，出笼后去掉姜片、葱白，加入食盐、鸡精调味即成。

Ingredients
2 pigeons, 50g Gastrodia tuber

Seasonings
2g salt, 2g chicken essence granules, 1g ground white pepper, 10g ginger (sliced), 15g scallion (white parts only), 15g Shaoxing cooking wine

Preparation
1. Clean the pigeons, and then blanch in water. Put the gastrodia tuber into a bowl, add water to the bowl, transfer the bowl to a steamer, and steam till the gastrodia tuber becomes soft. Slice the gastrodia tuber.
2. Mix water, pigeons, gastrodia tuber, ginger, scallion, Shaoxing cooking wine and ground white pepper in a large soup blow, cover and steam till the pigeon meat is soft. Remove and discard the ginger and scallion, and add salt and chicken essence granules.

色泽红亮，皮酥内嫩，味咸甜酸辣兼备，姜葱蒜味浓郁
brown color; tender eggs with crispy coating; a mixture of salty, sweet, sour and hot taste; strong aroma of garlic, scallion and ginger

鱼香虎皮鸽蛋
Tiger-skin Pigeon Eggs in Fish-Flavor Sauce

[原　料]
鸽蛋12个　食用油1000克（约耗80克）

[调　料]
泡辣椒末40克　姜米10克　蒜米15克　葱花20克　食盐3克　料酒10克　白糖25克　醋20克　酱油7克　味精2克　鲜汤50克　水淀粉10克　蛋清淀粉糊50克

[制　作]
1. 鸽蛋入锅煮熟，去壳备用；将食盐、料酒、白糖、醋、酱油、味精、鲜汤、水淀粉兑成芡汁。
2. 锅中下食用油烧至160℃，将鸽蛋裹上一层蛋清淀粉糊，入锅中炸至色黄后捞出；待油温回升到180℃时，将鸽蛋放入再炸至皮酥起皱时捞出装盘。
3. 锅中下食用油烧至120℃，放入泡辣椒末、姜米、蒜米、葱花炒香，倒入芡汁，收汁亮油后浇淋在鸽蛋上即成。

Ingredients
12 pigeon eggs, 1000g cooking oil for deep-frying

Seasonings
40g pickled chilies (finely chopped), 10g ginger (finely chopped), 15g garlic (finely chopped), 20g scallion (finely chopped), 3g salt, 10g Shaoxing cooking wine, 25g sugar, 20g vinegar, 7g soy sauce, 2g MSG, 50g everyday stock, 10g cornstarch-water mixture, 50g mixture of egg white and cornstarch

Preparation
1. Boil the pigeon eggs till cooked through. Remove the shells and rinse. Mix the salt, Shaoxing cooking wine, suger, vinegar, soy sauce, MSG, stock and cornstarch-water mixture to make thickening sauce.
2. Heat oil in a wok to 160℃. Coat the pigeon eggs with the mixture of egg white and cornstarch, deep-fry till brown, and remove from the oil. Reheat the oil to 180℃, add the pigeon eggs and deep-fry till the coating becomes crispy and wrinkled. Transfer the eggs to a serving dish.
3. Heat oil in a wok to 120℃, blend in the pickled chilies, ginger, garlic and scallion, stir-fry till aromatic and then add the thickening sauce. Wait till the sauce becomes thick and lustrous. Pour the sauce over the eggs.

色泽红亮，香气浓郁，咸鲜微辣略甜，肥而不腻
bright reddish color; lingering aroma; a combination of salty, savoury and slightly pungent and sweet flavors; fatty but not greasy taste

回锅肉
Twice-Cooked Pork

[原 料]
带皮猪坐臀肉300克 蒜苗100克 食用油40克

[调 料]
食盐0.6克 郫县豆瓣18克 甜面酱10克 酱油5克 白糖5克

[制 作]
1. 蒜苗洗净，斜切成约2.5厘米长的节。
2. 猪肉入锅煮熟，捞出晾冷后切成长约6厘米、厚约0.2厘米的薄片。
3. 锅中下食用油烧至150℃，放入肉片、食盐炒香并呈"灯盏窝"形，续下郫县豆瓣、甜面酱、酱油、白糖炒香上色，最后放入蒜苗炒断生起锅即成。

Ingredients
30g pork rump with skin attached, 100g leeks, 40g cooking oil

Seasonings
0.6g salt, 18g Pixian chili bean paste, 10g fermented flour paste, 5g soy sauce, 5g sugar

Preparation
1. Rinse the leeks and cut diagonally into 2.5cm lengths.
2. Boil the pork until cooked through, remove, cool and then cut into slices about 6cm long and 0.2cm thick.
3. Add the oil to the wok, heat to 150℃, add the pork and salt, and then stir-fry till the pork slices slightly curl up. Blend in Pixian chili bean paste, fermented flour paste, soy sauce and sugar, mix well and continue to stir-fry till aromatic. Add the leeks and stir-fry till al dente. Remove from the stove and transfer to a serving dish.

回锅肉 Twice-Cooked Pork

盐煎肉
Stir-Fried Pork with Leeks

色泽棕红，香味浓郁，咸鲜微辣
bright brown color; aromatic flavor; salty, delicate and slightly hot taste

[原 料]
猪腿肉200克 蒜苗50克 食用油50克

[调 料]
郫县豆瓣30克 豆豉8克 酱油4克 食盐1克 味精2克

[制 作]
1. 猪腿肉切成片；蒜苗切成节。
2. 锅置旺火上，下食用油烧热，下猪肉片、食盐反复煸炒至出油，续下郫县豆瓣、豆豉炒香，再入酱油、味精炒匀，最后加蒜苗炒断生起锅即成。

Ingredients
200g pork leg meat, 50g baby leeks, 50g cooking oil

Seasonings
30g Pixian chili bean paste, 8g fermented soy beans, 4g soy sauce, 1g salt, 2g MSG

Preparation
1. Cut the pork into slices. Cut the leeks into sections.
2. Heat the oil in a wok over a high flame. Add the pork slices and salt, stir-fry to evaporate the water till aromatic. Add the Pixian chili bean paste, fermented soy beans, soy sauce and MSG, and continue to stir-fry to blend well. Add the leeks, and stir-fry till al dente. Remove from the stove and transfer to a serving dish.

汤汁乳白，猪肘酥软，原汁原味
milky soup; tender and glutinous knuckle

东坡肘子
Dongpo Pork Knuckle

[原　料]
猪肘子750克　雪豆130克

[调　料]
姜片15克　葱段30克　料酒50克　食盐5克　酱油20克

[制　作]
1. 猪肘放入汤锅煮透，捞出后剔去肘骨，再放入垫有猪骨的砂锅内，入姜片、葱段、料酒、雪豆后加盖，移小火上煨炖3小时。
2. 食用时加食盐调味，也可用酱油味汁蘸食。

Ingredients
750g pork knuckle, 130g white haricot beans,

Seasonings
15g ginger (sliced), 30g scallion (cut into sections), 50g Shaoxing cooking wine, 5g salt, 20g soy sauce

Preparation
1. Boil the knuckle till soft, debone, and put into an earthen pot with pork bones in it. Add the ginger, scallion, Shaoxing cooking wine and haricot beans, cover and stew for three hours.
2. Add the salt before serving or serve with the soy sauce as the dipping sauce.

典故
东坡肘子为成都"味之腴"餐厅的当家名菜。相传炖肘子是根据苏东坡的烹制方法制作而成，故名。

Note
Dongpo Pork Knuckle is a specialty of the restaurant Weizhiyu in Chengdu. It is said that this recipe for stewing pork knuckle was devised by the famous poet Su Dongpo.

色泽红亮,肉香浓郁,肥而不腻
bright brown color; aromatic smell; fatty but not greasy pork

红烧肉
Red-Braised Pork Belly

[原 料]
带皮猪五花肉400克 食用油70克

[调 料]
八角4克 花椒1克 姜片5克 葱节15克 食盐6克 冰糖30克 料酒30克 糖色20克

[制 作]

1. 猪五花肉刮洗干净,入沸水锅内略煮后捞出,切成约3.3厘米见方的块。
2. 锅置中火上,入食用油烧至180℃,入肉块煸炒至水分干,加清水烧沸后移入砂锅内,加姜片、葱节、花椒、八角、冰糖、料酒、糖色、食盐,用小火烧至软熟后捡去姜片、葱节、花椒、八角不用,改用中火烧至汁浓即成。

Ingredients
400g pork belly with skin attached, 70g cooking oil

Seasonings
4g star anise, 1g Sichuan pepper, 5g ginger (sliced), 15g scallion (cut into sections), 6g salt, 30g rock sugar, 30g Shaoxing cooking wine, 20g caramel color

Preparation

1. Clean the pork thoroughly. Blanch in boiling water, remove and cut into 3.3cm^3 cubes.
2. Heat oil in a wok to 180℃ over a medium-high flame, slide in the pork cubes, stir-fry to vaporize the water in the pork. Add water, bring to a boil and then transfer to an earthen pot. Add ginger, scallion, Sichuan pepper, star anise, rock sugar, Shaoxing cooking wine, caramel color, and salt, and simmer over a low flame till the pork is soft and cooked through. Remove the ginger, scallion, Sichuan pepper, star anise and braise over a medium flame till the sauce thickens. Remove from the fire and transfer to a serving dish.

肉质软糯，味道浓厚

soft and tender meat; a diversity of ingredients and a medley of taste

坛子肉
Stewed Meat in an Earthen Pot

[原　料]

猪肘1000克　鸡肉500克　鸭肉500克　火腿50克　干墨鱼50克　水发海参150克　金钩50克　干贝50克　冬笋100克　口蘑30克　鸡蛋5个　猪骨1000克　食用油500克（约耗20克）

[调　料]

食盐8克　糖色20克　料酒20克　姜片15克　葱段25克　干淀粉20克

[制　作]

1. 猪肘、鸡肉、鸭肉分别切成4大块后汆水；火腿、水发海参、冬笋切片；鸡蛋煮熟去壳，粘上干淀粉，入180℃的食用油中炸至色金黄时捞出；干墨鱼用水泡软。
2. 猪骨置于陶质坛底，依次放入清水、料酒、姜片、葱段、食盐、糖色、火腿、猪肘、鸡肉、鸭肉、冬笋、口蘑、干贝、墨鱼、金钩，用皮纸封严坛口，置木炭火（微火）上煨4～5小时，再入水发海参、鸡蛋煨30分钟即成。

Ingredients

1000g pork knuckle, 500g chicken, 500g duck, 50g ham, 50g dried cuttlefish, 150g water-soaked sea cucumber, 50g dried and shelled shrimps, 50g dried scallops, 100g winter bamboo shoots, 30g St. George's mushrrooms, 5 eggs, 1000g pork bones, 500g cooking oil for deep-frying

Seasonings

8g salt, 20g caramel color, 20g Shaoxing cooking wine, 15g ginger (sliced), 25g scallion (cut into sections), 20g cornstarch,

Preparation

1. Chop the pork knuckle, chicken and duck respectively into four chunks and blanch. Slice the ham, water-soaked sea cucumber and winter bamboo shoots. Boil the eggs, remove the shells, coat with the cornstarch, deep-fry in 180℃ oil till golden brown, and remove. Soak the dried cuttlefish in water till soft.
2. Lay the pork bones on the bottom of an earthen pot, then add in turn water, Shaoxing cooking wine, ginger, scallion, salt, caramel color, ham, pork knuckle, chicken, duck, bamboo shoots, mushrooms, scallops, cuttlefish and shrimps. Cover the pot with brown paper and simmer for 4 to 5 hours over charcoal fire (low flames). Add the water-soaked sea cucumber and the eggs, then simmer for 30 more minutes.

鱼香肉丝

Pork Slivers in Fish-Flavor Sauce

色泽红亮，肉丝细嫩，咸甜酸辣兼备，姜葱蒜味突出
bright and lustrous color; tender pork; a mixture of salty, sweet, sour and slightly hot taste; scrumptious aroma of garlic, ginger and scallion

鱼香肉丝
Pork Slivers in Fish-Flavor Sauce

[原 料]
猪肉200克 青笋75克 水发木耳20克 食用油70克

[调料A]
食盐1克 料酒5克 水淀粉10克

[调料B]
泡辣椒末25克 姜米7克 蒜米14克 葱花20克

[调料C]
食盐1克 白糖10克 酱油6克 醋12克 料酒5克 味精1克 水淀粉15克 鲜汤25克

[制 作]
1. 猪肉切成二粗丝（长约8厘米、粗约0.3厘米的丝），与调味料A拌匀。
2. 青笋切成二粗丝；水发木耳切成细丝。
3. 锅中下食用油烧至170℃，放入肉丝炒熟，再放入调料B炒香，加入青笋丝及木耳丝，倒入调料C的混合液，收汁入味后起锅即成。

Ingredients
200g pork, 75g asparagus lettuce, 20g water-soaked Jew's ear, 70g cooking oil

Seasonings A
1g salt, 5g Shaoxing cooking wine, 10g cornstarch-water mixture

Seasonings B
25g pickled chilies (finely chopped), 7g ginger (finely chopped), 14g garlic (finely chopped), 20g scallion (finely chopped)

Seasonings C
1g salt, 10g sugar, 6g soy sauce, 12g vinegar, 5g Shaoxing cooking wine, 1g MSG, 15g cornstarch-water mixture, 25g everyday stock

Preparation
1. Cut the pork into slivers about 8cm in length and 0.3cm thick), mix with Seasonings A and marinate.
2. Cut the asparagus lettuce into slivers about 10cm long and 0.3cm thick. Cut the Jew's ear into strips.
3. Bring the oil to 170 ℃ in a wok, add pork slivers and stir-fry until cooked through. Add Seasonings B, sauté to bring out the aroma, and then add the asparagus lettuce and Jew's ear. Pour the mixture of Seasonings C into the wok, wait till the sauce becomes thick, and then transfer to a serving dish.

色泽棕黄，肉质细嫩，酱香浓郁，且带有葱香
Stir-Fried Pork Slivers with Fermented Flour Paste

酱肉丝

Stir-Fried Pork Slivers with Fermented Flour Paste

[原 料]
猪瘦肉150克 葱75克 食用油75克

[调 料]
食盐1克 料酒15克 水淀粉15克 甜面酱10克 味精3克 白糖3克 酱油5克 鲜汤30克

[制 作]
1. 猪瘦肉切成长约10厘米、粗约0.3厘米的丝，与食盐、料酒、水淀粉拌匀；葱白切成细丝。
2. 味精、白糖、酱油、料酒、鲜汤混合调成芡汁。
3. 锅置火上，入食用油烧至170℃，下肉丝炒散，再下甜面酱炒匀，倒入芡汁，收汁亮油后起锅装盘，放上葱丝即成。

Ingredients
150g lean pork, 75g scallion, 75g cooking oil

Seasonings
1g salt, 15g Shaoxing cooking wine, 15g cornstarch-water mixture, 10g fermented flour paste, 3g MSG, 3g sugar, 5g soy sauce, 30g everyday stock

Preparation
1. Cut the pork into slivers about 10cm long and 0.3cm thick, and blend evenly with the salt, 5g cooking wine, and corn-starch mixture. Cut the scallion white into slivers.
2. Mix the MSG, sugar, soy sauce, 10g cooking wine and stock to make the thickening sauce.
3. Heat oil in a wok to 170℃, slide in the pork slivers and stir-fry to separate them. Add the fermented flour paste, stir well and pour in the thickening sauce. Remove from the heat when the sauce becomes thick and lustrous. Transfer to a serving dish and sprinkle with scallion slivers.

青椒肉丝 Pork Slivers with Green Peppers

质地细嫩，清香微辣，咸鲜味浓
tender pork; fragrant smell; delicate and slightly hot taste

青椒肉丝
Pork Slivers with Green Peppers

[原　料]

猪肉200克　青椒100克　食用油75克

[调　料]

食盐2克　味精2克　酱油5克　鲜汤15克　水淀粉20克

[制　作]

1. 猪肉切成二粗丝，加食盐、水淀粉拌匀；青椒去籽后切成二粗丝；将食盐、酱油、味精、水淀粉、鲜汤调成芡汁。
2. 锅置旺火上，下食用油烧至170℃，下肉丝炒散，放青椒炒匀，加入芡汁，翻炒至汁浓亮油时起锅装盘即成。

Ingredients

200g pork, 100g green peppers, 75g cooking oil

Seasonings

2g salt, 2g MSG, 5g soy sauce, 15g everyday stock, 20g cornstarch-water mixture

Preparation

1. Cut the pork into slivers about 8cm long and 0.3cm thick, add salt and cornstarch-water mixture, and mix well. Core the green peppers and cut into slivers. Mix salt, soy sauce, MSG, cornstarch-water mixture and stock to make thickening sauce.
2. Heat oil in a wok to 170℃, add pork slivers and stir-fry so that they separate. Blend in the green peppers, stir well, pour in the thickening sauce and continue to stir till the sauce becomes thick and lustrous. Remove from the heat and transfer to a serving dish.

锅巴金黄酥脆，肉片细嫩，咸鲜甜酸

golden, crispy rice crust; tender pork; a mixture of salty, sour and sweet taste

锅巴肉片

Sliced Pork with Sizzling Rice Crust

[原 料]

锅巴150克 猪肉片100克 蘑菇片20克 水发兰片30克 鲜菜心30克 食用油1500克（约耗80克）

[调料A]

食盐1克 料酒5克 水淀粉10克

[调料B]

泡辣椒节15克 姜片5克 蒜片6克 葱节10克

[调料C]

食盐3克 酱油10克 白糖25克 醋25克 味精1克 水淀粉20克 鲜汤250克

[制 作]

1. 猪肉片与调料A拌匀。
2. 锅中下食用油50克烧至150℃，放入肉片炒断生，续下蘑菇片、水发兰片、鲜菜心及调料B、调料C，待汤汁变稠时起锅，装入大汤碗内。
3. 锅中下食用油烧至200℃，放入锅巴炸至色黄、酥脆时捞出装入大凹盘中，将肉片味汁淋在锅巴上即成。

Ingredients

150g rice crust, 100g pork (sliced), 20g fresh mushrooms (sliced), 30g water-soaked bamboo shoot slices, 30g tender leafy vegetables, 1500g cooking oil for deep-frying, 250g everyday stock

Seasonings A

1g salt, 5g Shaoxing cooking wine, 10g cornstarch-water mixture,

Seasonings B

15g pickled chilies (cut into sections), 5g ginger (sliced), 6g garlic (sliced), 10g scallion (chopped into sections)

Seasonings C

3g salt, 10g soy sauce, 25g sugar, 25g vinegar, 1g MSG, 20g cornstarch-water mixture,

Preparation

1. Mix the pork with Seasonings A evenly.
2. Heat 50g oil in a wok to 150℃, and sir-fry the pork slices till just cooked. Add sliced fresh mushrooms, water-soaked bamboo shoot slices and tenden leafy veget-ables and stir-fry until cooked through. Give Seasoning C a stir and tip into the wok. Wait until the soup thickens, remove from the heat and transfer to a large soup bowl.
3. Heat oil in a wok to 200℃, add the rice crust and deep-fry till it becomes crispy and golden brown. Remove and transfer to a deep serving dish. Then pour the pork and sauce in the soup bowl over the rice crust.

粉蒸肉
Steamed Pork Belly with Rice Flour

色泽红亮，肉质软糯，咸鲜略辣，入口回甜，香浓醇厚
bright color; soft and glutinous pork; lingering aroma; salty, slightly spicy and sweet taste

粉蒸肉
Steamed Pork Belly with Rice Flour

[原 料]
带皮猪五花肉200克 鲜豌豆75克 大米粉50克

[调 料]
郫县豆瓣25克 豆腐乳汁10克 姜米4克 葱青叶5克 花椒1克 食盐2克 味精4克 酱油5克 料酒10克 醪糟汁15克 糖色10克 鲜汤50克

[制 作]
1．猪五花肉切成长约10厘米、厚约0.3厘米的片；葱青叶、花椒铡细成粗椒麻糊。
2．肉片入盆内，加郫县豆瓣、粗椒麻糊、豆腐乳汁、姜米、食盐、味精、酱油、料酒、醪糟汁、糖色拌匀，再加入大米粉、鲜汤拌匀，静置15分钟后摆入蒸碗，再放上鲜豌豆，入笼内用旺火沸水蒸约2小时至肉软熟时出笼，翻扣在盘内即成。

Ingredients
200g pork belly with skin attached, 75g green peas, 50g rice flour

Seasonings
25g Pixian chili bean paste, 10g brine of fermented tofu, 4g ginger (finely chopped), 5g scallion (green parts only), 1g Sichuan pepper, 2g salt, 4g MSG, 5g soy sauce, 10g Shaoxing cooking wine, 15g fermented glutinous rice wine, 10g caramel color, 50g everyday stock

Preparation
1. Cut the pork into slices about 10cm long and 0.3cm thick. Make Jiaoma paste with the scallion and Sichuan pepper.
2. Put the pork slices into a large bowl, add the Pixian chili bean paste, Jiaoma paste, brine of fermented tofu, ginger, salt, MSG, soy sauce, Shaoxing cooking wine, fermented glutinous rice wine and caramel color, then blend well. Add the rice flour and stock, mix well and marinate for 15 minutes. Transfer the mixture to a steaming bowl, lay some green peas on top, and transfer to a steamer. Steam over a high flame for about 2 hours. Remove from the steamer and turn the steaming bowl upside down to transfer the contents onto a serving dish.

色泽金黄，外酥内嫩，甜酸香浓
golden color; tender pork that is crispy on the outside; sweet and sour tastes

糖醋里脊
Sweet-and-Sour Pork Tenderloin

[原 料]
猪里脊肉200克 食用油1000克（约耗100克）

[调料A]
食盐1克 料酒10克 鸡蛋淀粉75克

[调料B]
姜米5克 蒜米7克 葱花20克

[调料C]
食盐1克 味精2克 白糖30克 酱油3克 醋20克 料酒10克 水淀粉20克 鲜汤180克 芝麻油5克

[制 作]
1. 猪里脊肉切成长约4厘米、宽约1厘米的条，加调料A拌匀。
2. 将调料C调匀成芡汁。
3. 锅置火上，下食用油烧至150℃，入肉条炸至成熟后捞出；待油温回升到180℃时，再放入肉条炸至色金黄、质酥香时捞出。
4. 锅内留油少许，放调料B炒香，倒入调味芡汁烧至浓稠，放入炸过的肉条裹上汁液即成。

Ingredients
200g pork tenderloin, 1000g cooking oil for deep-frying

Seasonings A
1g salt, 10g Shaoxing cooking wine, 75g mixture of cornstarch and egg

Seasonings B
5g ginger (finely chopped), 7g garlic (finely chopped), 20g scallion (finely chopped)

Seasonings C
1g salt, 2g MSG, 30g sugar, 3g soy sauce, 20g vinegar, 10g Shaoxing cooking wine, 20g cornstarch-water mixture, 180g everyday stock, 5g sesame oil

Preparation
1. Cut the pork tenderloin into strips about 4cm long and 1cm wide. Mix evenly with Seasonings A.
2. Mix Seasonings C to make thickening sauce.
3. Heat the oil in a wok to 150℃, deep-fry pork strips till cooked through and remove from the oil. Heat the oil to 180℃, return the strips to the oil and deep-fry till their surfaces become golden brown and crispy.
4. Leave some oil in the wok, add Seasonings B and stir-fry till aromatic. Add the thickening sauce, and braise till the sauce is thick. Slide in the pork strips, and stir to coat evenly.

色泽棕红，味道鲜美，软糯不腻
brown color; delicate, aromatic and savoury taste; soft and glutinous pork that is fatty but not greasy

[原　料]

带皮猪五花肉250克　芽菜100克　泡辣椒节10克　豆豉20克　食用油1000克(约耗20克)

[调料A]

食盐1克　酱油10克　糖色5克　胡椒粉1克

[调料B]

姜片5克　葱段7克

[制　作]

1．芽菜洗净后切成长约1厘米的节，与泡辣椒节、豆豉拌匀。
2．猪五花肉煮熟后捞出、晾干表皮水分，趁热在肉皮表面抹上糖色，放入200℃的食用油中炸至表皮起皱、色呈棕红时捞出，放入热汤中浸泡至皮回软，再切成长约10厘米、厚约0.4厘米的片。
3．将肉片摆放在蒸碗中，上面放芽菜节，倒入调料A的混合液，再放上调料B。
4．蒸碗入笼，用旺火蒸2小时，取出翻扣在盘内即成。

Ingredients

250g pork belly with skin attached, 100g yacai (preserved mustard stems), 10g pickled chilies (chopped into sections), 20g fermented soy beans, 1000g cooking oil for deep-frying,

Seasonings A

1g salt, 20g soy sauce, 5g caramel color, 1g ground white pepper

Seasonings B

5g ginger (sliced), 7g scallion (cut into sections)

Preparation

1. Wash the yacai thoroughly, chop into 1cm-long sections, and mix evenly with pickled chilies and fermented soy beans.
2. Boil the pork till cooked through, remove and drain. Smear the pork skin with caramel color while the pork is still hot, and deep-fry in 200℃ oil till the skin becomes wrinkled and brown. Soak the fried pork in the hot broth where the pork has been boiled till the skin becomes soft, and then cut into slices about 10cm long and 0.4cm thick.
3. Lay the slices into a steaming bowl, cover with yacai, pour the mixture of Seasonings A over yacai, and then lay Seasonings B on top.
4. Put the bowl into a steamer and steam over a high flame for 2 hours. Remove and turn the bowl upside down to transfer the contents onto a serving dish.

咸烧白

Steamed Pork with Salty Stuffing

甜烧白

Steamed Pork with Sweet Stuffing

肥而不腻，软糯润滑，香甜适口

savoury and sweet taste, glutinous pork that's fatty but not greasy

【原　料】

带皮猪五花肉500克　糯米100克　豆沙100克　食用油1000克（约耗20克）

【调　料】

红糖50克　猪油20克　白糖50克　糖色5克

【制　作】

1. 猪五花肉刮洗干净，入汤锅中煮熟后捞出，趁热在皮上抹一层糖色，晾干水分，放入油锅中炸至表皮起泡、上色后待用。
2. 糯米淘洗干净，放入沸水锅中煮至无硬心时捞出，趁热加入红糖、糖色、猪油拌匀。
3. 将上色后的猪肉切成长约8厘米、宽约3厘米、厚约0.4厘米的夹层片，每片肉中夹上豆沙，皮向下装入蒸碗，上面填入糯米，上笼蒸2小时，食用时翻扣装入盘内，撒上白糖即成。

Ingredients

500g pork belly with skin attached, 100g glutinous rice, 100g red bean paste, 1000g cooking oil for deep-frying

Seasonings

50g brown sugar, 20g lard, 50g sugar, 5g caramel color

Preparation

1. Clean the pork, boil in water till cooked through. Remove the pork, smear its skin with caramel color while it is still hot, and then leave to drain. Deep-fry the pork in the cooking oil till its skin bubbles and browns.
2. Boil the glutinous rice in water till soft, remove and then mix well with brown sugar, caramel color and lard.
3. Cut the pork into slices about 8cm long, 3cm wide and 0.4cm thick. Make one cut in each slice to sandwich the red bean paste. Lay the sandwiched slices into a steaming bowl, making sure that the skin is in contact with the base of the bowl. Cover the pork slices with glutinous rice steam for two hours, and remove. Turn the bowl upside down to transfer its contents onto a serving dish. Sprinkle with the sugar.

脆嫩爽口，清香浓郁，咸鲜味美
crispy and crunchy ingredients; fragrant smell; salty, appetizing and savory taste

火爆双脆
Crispy Quick-Fried Pork Tripe and Chicken Gizzards

[原 料]

猪肚头125克 鸡胗125克 玉兰片25克 菜心30克 食用油75克

[调 料]

泡辣椒节10克 姜片5克 蒜片8克 葱节10克 食盐2克 味精1克 料酒10克 水淀粉15克 芝麻油3克 鲜汤30克

[制 作]

1. 肚头剞上花纹，再切为约2厘米大小的菱形块；鸡胗去筋，剞上花纹，每个切成四块；将肚头、鸡胗用食盐、料酒、水淀粉拌匀；食盐、味精、料酒、水淀粉、芝麻油、鲜汤调成芡汁。
2. 锅置旺火上，下食用油烧至200℃，放入猪肚、鸡胗爆炒，续下泡辣椒节、姜片、蒜片、葱节炒香，再下玉兰片、菜心炒匀，调入芡汁，收汁亮油后起锅装盘即成。

Ingredients

125g pork tripe, 125g chicken gizzards, 25g bamboo shoot slices, 30g tender leafy vegetables, 75g cooking oil

Seasonings

10g pickled chilies (cut into sections), 5g ginger (sliced), 8g garlic (sliced), 10g scallion (cut into sections), 2g salt, 1g MSG, 10g Shaoxing cooking wine, 15g cornstarch-water mixture, 3g sesame oil, 30g everyday stock

Preparation

1. Lay the tripe flat on a chopping board, make parallel cuts across its surface, then make similar cuts perpendicular to the initial cuts. Cut the tripe into 2cm-long diamonds. Remove the tendon from the chicken gizzards, and cut each into four parts. Mix the trip, gizzards, salt, Shaoxing cooking wine and cornstarch-water mixture, and blend well. Mix salt, MSG, Shaoxing cooking wine, cornstarch-water mixture, sesame oil and stock to make thickening sauce.
2. Put a wok over a high flame, pour in the cooking oil and heat to 200℃. Stir-fry the tripe and gizzards, add the pickled chilies, ginger, garlic and scallion, then go on stir-frying till aromatic. Slide in the bamboo shoot slices and tender leafy vegetables, and stir well. Pour in the thickening sauce, wait till the sauce becomes thick and lustrous, and then transfer to a serving dish.

火爆腰花
Quick-Fried Pork Kidneys

色泽协调，成形美观，质地脆嫩
harmoniously arranged colors; exquisite kidney slices; tender and slightly crunchy texture

火爆腰花
Quick-Fried Pork Kidneys

[原　料]
猪腰250克 青笋75克 食用油75克

[调料A]
食盐1克 料酒5克 淀粉10克

[调料B]
食盐2克 白糖2克 味精1克 胡椒粉1克 酱油5克 料酒5克 鲜汤20克 水淀粉15克 芝麻油5克

[调料C]
姜片3克 蒜片5克 马耳朵形泡辣椒（斜切成2.5厘米长的节）15克 马耳朵形葱10克

[制　作]
1. 猪腰去筋膜，切成凤尾形（三刀一断），与调料A拌匀；青笋切为筷子条（长约4厘米、粗约0.6厘米）。
2. 将调料B调匀成芡汁。
3. 锅置旺火上，下食用油烧至180℃，入腰花炒至断生，续下调料C及青笋条炒出香味，倒入芡汁，待收汁亮油后起锅装盘即成。

Ingredients
250g pork kidneys, 75g asparagus lettuce, 75g cooking oil

Seasonings A
1g salt, 5g Shaoxing cooking wine, 10g cornstarch

Seasonings B
2g salt, 2g sugar, 1g MSG, 1g ground white pepper, 5g soy sauce, 5g Shaoxing cooking wine, 20g everyday stock, 15g cornstarch-water mixture, 5g sesame oil

Seasonings C
3g ginger (sliced), 5g garlic (sliced), 15g pickled chilies (cut diagonally into sections shaped like "horse ear" 2.5cm in length), 10g scallion (cut diagonally into sections shaped like "horse ear")

Preparation
1. Remove the tendon and membrane of the pork kidneys. Cut into slices and make two cuts into each slice. Mix the slices with Seasonings A. Cut the asparagus lettuce into slivers about 4cm long and 0.6cm thick.
2. Mix Seasonings B to make thickening sauce
3. Heat oil in a wok over a high flame to 180℃, add the kidney slices and stir-fry till just cooked. Add Seasonings C and asparagus lettuce, and then stir-fry till aromatic. Pour in the thickening sauce and blend well. Wait till the sauce becomes thick and lustrous, remove from the stove and transfer to a serving dish.

本味清鲜，味汁微辣
delicate soup; spicy dipping sauce

[原 料]
带皮猪后腿肉300克 去皮萝卜750克 食用油75克

[调 料]
姜片10克 葱段20克 花椒2克 郫县豆瓣30克 酱油50克 食盐5克 味精3克 芝麻油5克 胡椒粉1克

[制 作]
1．锅置火上，下食用油烧至120℃，入郫县豆瓣炒至油呈红色时起锅装入碗内，加酱油、味精、芝麻油调匀成味碟蘸料。
2．猪肉洗净入锅，加清水用旺火煮沸，放姜片、葱段、花椒，用小火煮至肉熟时捞出晾冷，切成皮肉相连，长约10厘米的薄片。
3．萝卜切成长约6厘米、宽约3厘米、厚约0.3厘米的片，放入煮肉的汤中煮至软熟，放入肉片、胡椒粉、食盐、味精煮沸后起锅，配上味碟蘸料即成。

Ingredients
300g pork hind leg meat with skin attached, 750g peeled radish, 75g oil

Seasonings
10g ginger (sliced), 20g scallion (cut into sections), 2g Sichuan pepper, 30g chili bean paste, 50g soy sauce, 5g salt, 3g MSG, 5g sesame oil, 1g ground white pepper

Preparation
1. Heat oil in a wok to 120℃, add the chili bean paste, stir-fry till the oil becomes reddish, and transfer to a bowl. Add the soy sauce, MSG and sesame oil to the bowl. Whisk to make dipping sauce.
2. Wash thoroughly the pork and put into a pot. Add water to the pot, and bring to a boil. Add the ginger, scallion and Sichuan pepper, then simmer till the pork is fully cooked. Remove, cool and cut into 10cm-long thin slices with skin on each slice.
3. Cut the radish into slices about 6cm long, 3cm wide and 0.3cm thick, and place into the pre-simmered soup. Boil the radish till soft. Add the pork slices, ground white pepper, salt and MSG, and then bring to a second boil. Remove from the heat, transfer to a serving bowl, and serve hot with the dipping sauce.

萝卜连锅汤
Pork Soup with Radish

香辣猪蹄
Hot-and-Spicy Pork Feet

色泽金红，香辣味浓，质地软韧
golden brown color; spicy and pungent taste; glutinous but tenacious pork feet

香辣猪蹄
Hot-and-Spicy Pork Feet

[原 料]
猪蹄100克 卤水2000克 食用油1000克（约耗50克）

[调 料]
干辣椒75克 青花椒20克 姜片10克 蒜片10克 葱段15克 香辣油30克 料酒10克 酱油5克 味精3克 芝麻油5克

[制 作]
1. 猪蹄初加工后先入沸水锅中汆一水，再放入卤水锅中卤制2～3小时捞出，晾冷后去掉大骨，斩成约4厘米见方的块。
2. 锅中下食用油烧至180℃，放入猪蹄炸香后捞出。
3. 锅中留油少许，放入干辣椒、青花椒、姜片、蒜片、葱段炒香，再下猪蹄、香辣油、料酒、酱油、味精、芝麻油翻炒入味，起锅装盘即成。

Ingredients
100g pork feet, 2000g Sichuan-style broth, 1000g cooking oil for deep-frying

Seasonings
75g dried chilies, 20g green Sichuan peppers, 10g ginger (sliced), 10g garlic (sliced), 15g scallion (cut into sections), 30g chili oil, 10g Shaoxing cooking wine, 5g soy sauce, 3g MSG, 5g sesame oil

Preparation
1. Clean the feet, and blanch in water. Transfer the feet to a pot containing the Sichuan-style broth and simmer for 2 to 3 hours. Remove the pork feet from the broth, cool and then remove the big bones. Cut the pork feet into 4cm³ cubes.
2. Heat the cooking oil in a wok to 180℃, and deep-fry the pork trotters till aromatic.
3. Drain off most of the oil, and stir-fry in the remaining oil the dried chilies, green Sichuan pepper, ginger, garlic, scallion to bring out the aroma. Add pork feet, chili oil, Shaoxing cooking wine, MSG and sesame oil, stir-fry till fragrant and transfer to a serving dish.

雪豆蹄花

Pork Feet Stew with White Haricot Beans

汤汁乳白，猪蹄软糯

milky soup; soft and glutinous pork trotters; delicate taste

[原 料]

猪蹄750克 雪豆150克

[调 料]

姜片15克 葱段30克 料酒50克 花椒2克 食盐5克 油酥豆瓣30克 芝麻油4克

[制 作]

1. 猪蹄入汤锅内煮透后捞出洗净，放入砂锅内，加姜片、葱段、料酒、花椒、雪豆，加盖烧沸，移小火上炖制3小时。
2. 食用时拣去姜片、葱段、花椒不用，加食盐调味，用油酥豆瓣、芝麻油调成的酱料蘸食即可。

Ingredients

750g pork trotters, 150g white haricot beans

Ingredients

15g ginger (sliced), 30g scallion (cut into section), 50g Shaoxing cooking wine, 2g Sichuan pepper, 5g salt, 30g stir-fried chili bean paste, 4g sesame oil

Preparation

1. Boil the pork trotters in a soup pot till cooked through, remove, rinse and transfer to an earthen pot. Add the ginger, scallion, Shaoxing cooking wine, Sichuan pepper, and white haricot beans. Bring to a boil and then simmer over a low flame for three hours.
2. Remove and discard the ginger, scallion and Sichuan pepper, and then add some salt while serving. Blend the fried chili bean paste and sesame oil to make the dipping sauce.

滑润适口,葱香浓郁,味道鲜美
smooth, scallion-flavored tendon; delicate and savoury taste

葱烧蹄筋
Braised Pork Tendon with Scallion

[原 料]

水发蹄筋750克 葱段100克 食用油50克

[调 料]

姜片15克 食盐3克 蚝油10克 酱油5克 胡椒粉1克 料酒10克 水淀粉15克 芝麻油5克 奶汤300克

[制 作]

1. 将发好的蹄筋切条待用。
2. 锅中下食用油烧热,入姜片、葱段炒香,续下奶汤、蚝油、酱油、食盐、胡椒粉、料酒、蹄筋烧至入味,用水淀粉勾芡,淋入芝麻油推匀,起锅装盘即成。

Ingredients

750g water-soaked pork tendon, 100g scallion (cut into sections), 50g cooking oil

Seasonings

15g ginger (sliced), 3g salt, 10g oyster sauce, 5g soy sauce, 1g ground white pepper, 10g Shaoxing cooking wine, 15g cornstarch-water mixture, 5g sesame oil, 300g milky stock

Preparation

1. Cut the pork tendon into strips.
2. Heat oil in a wok, add the ginger and scallion, and stir-fry to bring out the aroma. Add the milky stock, oyster sauce, soy sauce, salt, ground white pepper, Shaoxing cooking wine and pork tendon, simmer so that the tendon absorbs the rich flavors, and then add the cornstarch-water mixture to thicken the sauce. Tip in the sesame oil, stir well and transfer to a serving dish.

脆嫩爽口，咸鲜微辣
beef spareribs that are crispy on the outside and tender within; salty and slightly hot taste

川椒牛仔骨
Sichuan-Style Pepper-Flavored Beef Spareribs

[原　料]

牛仔骨400克 红甜椒50克 青尖椒50克 洋葱50克 食用油1000克（约耗70克）

[调料A]

食盐1克 料酒10克 蛋清30克 干淀粉30克

[调料B]

食盐2克 酱油10克 料酒10克 黑胡椒粉3克 味精2克 芝麻油6克 辣椒油20克 鲜汤100克 水淀粉20克

[制　作]

1. 牛仔骨切成长约10厘米、宽约5厘米、厚约1厘米的片，用调料A码味1小时。
2. 红甜椒、青尖椒、洋葱切成丁。
3. 锅中下食用油烧至130℃，放入牛仔骨炸熟，装入盘中；锅中留油少许，放入红甜椒、青尖椒、洋葱炒断生，放入调料B，收汁浓稠后浇淋在牛仔骨上即成。

Ingredients

400g beef spareribs, 50g red bell peppers, 50g green chili peppers, 50g onion, 1000g cooking oil for deep-frying

Seasonings A

1g salt, 10g Shaoxing cooking wine, 30g eggwhite, 30g cornstarch

Seasonings B

2g salt, 10g soy sauce, 10g Shaoxing cooking wine, 3g ground black pepper, 2g MSG, 6g sesame oil, 20g chili oil, 100g everyday stock, 20g cornstarch-water mixture

Preparation

1. Cut the spareribs into chunks about 10cm long, 5cm wide and 1cm thick. Marinate with Seasonings A for an hour.
2. Cut the red bell peppers, green chili peppers and onion into strips.
3. Heat oil in a wok to 130℃, add the sparerib chunks and deep-fry till cooked through. Transfer to a plate. Leave a little oil in the wok, stir-fry the red bell peppers, green chili peppers and onion till al dente, and add Seasonings B. Wait till the sauce thickens and pour over the spareribs.

Red-Braised Water Buffalo Scalp
红烧牛头方

色泽棕红发亮，质地软糯，滋味醇浓
a deep, glossy red-brown color; glutinous skin cubes; lingering aroma

[原 料]

水牛头皮1000克　火腿片50克　鸡翅200克　鸡骨500克　菜心250克　化鸡油20克

[调 料]

食盐6克　姜15克　葱段20克　料酒50克　糖色20克　胡椒粉2克　鸡汤2000克　芝麻油5克

[制 作]

1. 水牛头皮用明火燎烧后去尽残毛和粗皮，用清水煮至软熟后取出，切成大方块成牛头方，用沸水氽煮3次；鸡翅、鸡骨氽水后备用。
2. 锅置火上，依次放入鸡骨、鸡翅、牛头方、火腿片、姜、葱段，掺入鸡汤，加入食盐、料酒、胡椒粉、糖色烧沸，去掉浮沫，用小火煨至牛头方炖软。
3. 菜心入沸水锅中氽断生，入盘垫底；将牛头方、火腿片出锅放在菜心上，再将锅内的原汁用旺火收浓，加味精、芝麻油、化鸡油推匀，浇淋在牛头方上即成。

Ingredients

1000g water buffalo scalp, 50g ham (sliced), 200g chicken wings, 500g chicken bones, 250g tender leafy vegetables, 20g chicke oil

Seasonings

6g salt, 15g ginger, 20g scallion (cut into sections), 50g Shaoxing cooking wine, 20g caramel color, 2g ground white pepper, 2000g chicken stock, 5g sesame oil

Preparation

1. Scald the buffalo scalp, remove any hair and rough skin and boil in water till cooked through. Remove, cut into chunks and blanch three times. Blanch the chicken wings and bones.
2. Heat a wok over a flame, add the chicken bones, chicken wings, buffalo scalp, ham, ginger, scallion, chicken stock, salt, Shaoxing cooking wine, ground white pepper and caramel color, and then bring to a boil. Skim and then simmer till the scalp pieces are tender.
3. Blanch the vegetables and transfer to a serving dish. Remove the buffalo scalp and ham from the wok, and lay on top of the vegetables. Heat the remaining soup in the wok over a high flame till it becomes thick. Add the MSG, sesame oil and chicken oil, mix well and pour over the buffalo head skin.

色泽棕红油亮，质地酥软化渣，麻辣鲜香浓郁
brown color; crispy, tender beef; spicy and aromatic flavor

干煸牛肉丝
Dry-Fried Beef Slivers

[原 料]
瘦牛肉300克 芹菜100克 食用油150克

[调 料]
郫县豆瓣30克 姜丝10克 蒜丝10克 花椒粉1克 食盐1克 酱油10克 芝麻油10克 味精1克

[制 作]
1. 牛肉切成长约8厘米、粗约0.4厘米的丝；芹菜切成长约4厘米的段。
2. 锅置旺火上，下食用油烧至180℃，加入牛肉丝煸炒至水气将干时放入食盐、姜丝、蒜丝、郫县豆瓣炒香，续下酱油、芹菜炒断生，再入芝麻油、味精炒匀，起锅装盘，撒上花椒粉即成。

Ingredients
300g lean beef, 100g celery, 150g cooking oil

Seasonings
30g Pixian chili bean paste, 10g ginger (shredded), 10g garlic (shredded), 1g ground roasted Sichuan pepper, 1g salt, 10g soy sauce, 10g sesame oil, 1g MSG

Preparation
1. Cut the beef into slivers about 8cm long and 0.4cm thick. Chop the celery into 4cm lengths.
2. Heat oil in a wok over a high flame to 180℃, slide in the beef slivers and stir-fry till they have almost completely lost their water content. Add the salt, ginger, garlic and chili bean paste, and stir-fry to bring out the fragrance. Add the soy sauce and celery, and stir-fry till al dente. Season with sesame oil and MSG. Blend well, transfer to a serving dish and sprinkle with ground roasted Sichuan pepper.

色泽红亮，牛肉滑嫩，蔬菜清香，咸鲜麻辣
bright and fiery color; tender and smooth beef; fragrant vegetables; salty, savoury, numbing and spicy taste

水煮牛肉
Boiled Beef in Chili Sauce

[原　料]

牛肉150克　蒜苗50克　芹菜70克　莴笋尖120克　刀口辣椒（干辣椒20克、花椒10克炒香剁碎）30克　食用油100克

[调料A]

食盐1克　料酒5克　水淀粉30克

[调料B]

郫县豆瓣25克

[调料C]

食盐1克　味精1克　酱油8克　料酒5克　鲜汤350克

[制　作]

1. 蒜苗、芹菜切成长约10厘米的段；莴笋尖切成片；三料混合后入锅，加少许食盐炒熟，装碗垫底。
2. 牛肉切成片，用调料A拌匀。
3. 锅中下食用油烧至120℃，放入调料B炒香后下调料C和牛肉片煮熟，倒入碗中，撒上刀口辣椒，浇上少许热油即成。

Ingredients

150g beef, 50g baby leeks, 70g celery, 120 asparagus lettuce tips (the leafy parts), 30g blade-minced chilies (mix 20g dried chilies with 10g Sichuan pepper, sautéed and minced), 100g cooking oil

Seasonings A

1g salt, 5g Shaoxing cooking wine, 30g cornstarch-water mixture

Seasoning B

25g chili bean paste

Seasonings C

1g salt, 1g MSG, 8g soy sauce, 5g Shaoxing cooking wine, 350g everyday stock

Preparation

1. Chop the leeks and celery into 10cm lengths, slice asparagus lettuce tips, mix them and sautéed in a wok till cooked through. Add some salt and transfer to a bowl.
2. Slice the beef, mix evenly with Seasonings A.
3. Add oil to a wok, heat to 120℃, add Seasong B and stir-fry to bring out the aroma. Add the stock, Seasonings C and beef slices. Boil till the beef is cooked through. Pour the contents in the wok into the bowl, sprinkle with the blade-minced chilies, and pour some hot oil onto the chilies.

色泽棕红，肉质软嫩，咸鲜香辣微麻
brown color; tender beef and melting rice flour; salty, spicy, hot and slightly tingling taste

[原 料]
牛肉150克 米粉20克

[调 料] A
郫县豆瓣10克 豆豉2克 红酱油5克 粗椒麻糊3克 姜米2克 料酒5 白糖1克 菜油20克 豆腐乳汁4克 胡椒粉0.5克

[调 料] B
蒜泥5克 干辣椒粉5克 花椒粉1克 葱花6克 香菜末6克

[制 作]
1. 牛肉切成约5厘米大小的薄片，放入调料A拌匀，再加米粉、适量清水调匀，待米粉吸水后装入特制小竹笼。
2. 将小竹笼放入笼屉内，用旺火蒸30分钟后取出，加调料B即成。

Ingredients
150g beef, 20g rice flour for steaming meat

Seasonings A
10g Pixian chili bean paste, 2g fermented soy beans, 5g soy sauce, 3g Jiaoma paste, 2g ginger (finely-chopped), 5g Shaoxing cooking wine, 1g sugar, 20g rapeseed oil, 4g brine of fermented tofu, 0.5g ground white pepper

Seasonings B,
5g garlic (finely chopped), 5g ground chilies, 1g ground roasted Sichuan pepper, 6g scallion (finely chopped), 6g coriander leaves (finely chopped)

Preparation
1. Cut the beef into 5cm-long slices, mix with Seasonings A and blend well. Add the rice four and water, mix well and transfer to a specially-made small bamboo steamer.
2. Transfer the bamboo steamer to a steaming pot, and steam over a high flame for 30 minutes. Remove and add Seasonings B.

小笼蒸牛肉
Steamed Beef in a Small Bamboo Steamer

色泽红亮，软糯醇香，咸鲜微辣

tender, savoury and juicy beef; salty, scrumptious and spicy taste

竹笋烧牛肉
Braised Beef with Bamboo Shoots

[原　料]
牛肋条肉400克　竹笋200克　食用油60克

[调　料]
葱段5克　姜片5克　香料25克　花椒2克　郫县豆瓣25克　料酒20克　白糖5克　味精3克　食盐4克　香菜5克　水淀粉15克　鲜汤1000克

[制　作]
1. 竹笋切成长约2厘米的段；牛肋条肉切成约3厘米见方的块，入沸水中煮3分钟后捞出。
2. 锅置旺火上，下食用油烧至120℃，入郫县豆瓣、姜片、葱段、香料、花椒炒香，加鲜汤、牛肉，用旺火烧沸后加料酒、食盐、白糖，改用小火煨2小时，再放入竹笋煮30分钟，加味精，用水淀粉勾芡，起锅后撒上香菜即成。

Ingredients
400g intercostals meat of beef ribs, 200g bamboo shoots, 60g cooking oil

Seasonings
5g scallion (cut into sections), 5g ginger (sliced), 25g mixed herbal spices, 2g Sichuan pepper, 25g Pixian chili bean paste, 20g Shaoxing cooking wine, 5g sugar, 3g MSG, 4g salt, 5g coriander leaves, 15g cornstarch-water mixture, 1000g everyday stock

Preparation
1. Cut the bamboo shoots into 2cm-long sections. Dice the beef into 3cm³ cubes and boil in water for 3 minutes.
2. Heat oil in a wok to 120℃, add Pixian chili bean paste, ginger, scallion, herbal spices and Sichuan pepper, and stir-fry till aromatic. Add the stock, bring to a boil over a high flame, add beef, Shaoxing cooking wine, salt and sugar, and then simmber for 2 hours. Add the bamboo shoots, boil for 30 minutes, and then add MSG and the cornstarch-water mixture to thicken the sauce. Sprinkle with cilantro and transfer to a serving dish.

色泽红亮，香辣浓郁
bright and lustrous color; pungent and aromatic taste

香辣肥牛
Hot-and-Spicy Beef

[原　料]
肥牛肉300克　洋葱50克　红甜椒50克　青椒50克　香菜50克　食用油1000克（约耗70克）

[调　料]
辣椒粉20克　花椒粉3克　食盐3克　白糖3克　料酒10克　味精2克　芝麻油5克　花椒油5克　辣椒油10克

[制　作]
1．洋葱、红甜椒、青椒切成丝；香菜切成节。
2．肥牛肉切成薄片，放入150℃的食用油中炸熟后捞出。
3．锅中留油少许，放入辣椒粉、花椒粉炒香，续下洋葱、红甜椒、青椒、香菜、肥牛肉炒熟出香，再入食盐、白糖、料酒、味精、芝麻油、花椒油、辣椒油炒匀，起锅装盘即成。

Ingredients
300g beef, 50g onions, 50g red bell peppers, 50g green peppers, 50g coriander leaves, 1000g cooking oil for deep-frying

Seasonings
20g ground chilies, 3g ground roasted Sichuan pepper, 3g salt, 3g sugar, 10g Shaoxing cooking wine, 2g MSG, 5g sesame oil, 5g Sichuan pepper oil, 10g chili oil

Preparation
1. Cut the onions, red bell peppers and green peppers into strips. Cut the coriander into sections.
2. Cut the beef into thin slices, deep-fry in 150℃ oil till cooked through, and remove.
3. Drain off most of the oil, stir-fry in the remaining oil the ground chilies and ground roasted Sichuan pepper to bring out the aroma, add the onions, red bell peppers, green peppers, beef and stir-fry till aromatic. Add the salt, sugar, Shaoxing cooking wine, MSG, sesame oil, Sichuan pepper oil, chili oil, blend well and transfer to a serving dish.

香辣肥牛
Hot-and-Spicy Beef

207

清香味浓，肥牛鲜嫩
fragrant smell; tender and savoury beef

藤椒肥牛
Beef with Green Sichuan Pepper

[原 料]
肥牛片250克 金针菇200克

[调 料]
藤椒油10克 豉油50克 食盐3克 味精2克 酱油10克 料酒20克 鲜汤50克 鲜藤椒10克 鲜青辣椒圈15克

[制 作]
1. 金针菇入沸水煮1分钟后捞出，置于盘中垫底。
2. 锅置火上，入清水、料酒、食盐烧沸，将肥牛片入锅煮20秒后捞出，沥干水分，摆在金针菇上。
3. 豉油、食盐、味精、酱油、藤椒油、鲜汤入锅烧沸，倒入盛菜的盘中，再将鲜藤椒、鲜青辣椒圈放在肥牛片上，淋上180℃的热油即成。

Ingredients
250g beef slices, 200g enoki mushrooms

Seasonings
10g green Sichuan pepper oil, 50g fermented-soy-bean-flavored oil, 3g salt, 2g MSG, 10g soy sauce, 20g Shaoxing cooking wine, 50g everyday stock, 10g green Sichuan pepper, 15g green chili peppers (chopped into thin rings)

Preparation
1. Blanch the enoki mushrooms, remove, drain and lay onto a serving dish.
2. Add water, Shaoxing cooking wine and salt to a wok, bring to a boil, add beef slices and boil for 20 seconds. Remove the beef, drain and lay on top of the enoki mushrooms.
3. Add the fermented-soy-bean-flavored oil, salt, MSG, soy sauce, green Sichuan pepper oil and stock to a wok and bring to a boil. Pour the mixture over the beef, garnish with green Sichuan pepper and green chili pepper rings on the top, and then drizzle with 180℃ oil.

汁红油亮，兔肉细嫩，咸鲜带辣
bright color and lustrous oil; tender and soft rabbit; spicy and appetizing taste

鲜椒仔兔
Rabbits with Chili Peppers

[原料]
净仔兔肉400克 鲜青尖椒50克 鲜红尖椒50克 食用油1000克（约耗60克）

[调料A]
食盐1克 料酒10克 蛋清淀粉80克

[调料B]
辣豆瓣30克 花椒6克 姜片10克 蒜片10克 葱段25克 食盐1克 料酒10克 味精2克 芝麻油5克 辣椒油20克

[制作]
1. 鲜青尖椒、鲜红尖椒切成长约3厘米的段；兔肉切成约3厘米见方的块，与调料A拌匀。
2. 锅中下食用油烧至100℃，放入兔块滑熟后捞出。
3. 锅中留油少许，下辣豆瓣、花椒、鲜青尖椒、鲜红尖椒、姜片、蒜片、葱段炒香，再入兔丁、食盐、料酒、味精、芝麻油、辣椒油炒匀，起锅装盘即成。

Ingredients
400g boned rabbit meat, 50g green chili peppers, 50g red chili peppers, 1000g cooking oil for deep-frying

Seasonings A
1g salt, 10g Shaoxing cooking wine, 80g mixture of egg white and cornstarch

Seasonings B
30g chili bean paste, 6g Sichuan pepper, 10g ginger (sliced), 10g garlic (sliced), 25g scallion (cut into sections), 1g salt, 10g Shaoxing cooking wine, 2g MSG, 5g sesame oil, 20g chili oil

Preparation
1. Cut the green chili peppers and red chili peppers into 3cm-long sections. Dice the rabbit into $3cm^3$ cubes and mix with Seasonings A.
2. Heat the oil in a wok to 100℃, add the rabbit cubes, and deep-fry till cooked through.
3. Drain off most of the oil and stir-fry in the remaining oil the bean paste, Sichuan pepper, green chili peppers, red chili peppers, ginger, garlic and sallion to bring out the aroma. Add the rabbit cubes, salt, Shaoxing cooking wine, MSG, sesame oil and chili oil, stir well and transfer to a serving dish.

Rabbits with Chili Peppers
鲜椒仔兔

色泽红亮，麻、辣、烫、咸、鲜、酥、嫩，豆腐形态完整
lustrous reddish color; beautifully-arranged tofu; a combination of numbing, pugent, hot, salty, savoury, crispy and tender taste

麻婆豆腐
Mapo Tofu

典故

清朝同治年间，成都"陈兴盛饭铺"的店主之妻陈氏精心烹制出来的豆腐，浓厚入味，麻辣可口，风味独特，人们为区别于其他烧豆腐，见她脸上有麻痕，便戏称她烹制的豆腐为"麻婆豆腐"。

Note

During the reign of Emperor Tongzhi of the Qing Dynasty, the wife of the owner of Chen Xing Sheng Restaurant invented a way to cook tofu, which featured distinct spicy flavor. To distinguish it from other braised tofu, people named the dish Mapo Tofu, which in Chinese means pock-faced granny on account of the fact that there were pocks on her face.

[原　料]
豆腐300克　牛肉臊末60克　蒜苗节20克　食用油80克

[调料A]
郫县豆瓣25克　辣椒粉10克　豆豉6克

[调料B]
食盐3克　酱油5克　味精1克

[调料C]
花椒粉1克　鲜汤200克　水淀粉30克

[制　作]
1. 豆腐切成约1.8厘米见方的块，放入加有食盐的水中煮沸后捞出，用清水浸泡备用。
2. 锅中下食用油烧至120℃，放入调料A炒香，掺鲜汤，放豆腐、牛肉臊末和调料B，用小火烧2～3分钟，放入蒜苗节，用水淀粉收浓汁水后起锅装碗，撒上花椒粉即成。

Ingredients
300g tofu, 60g stir-fried beef mince, 20g baby leeks (chopped into sections), 80g cooking oil

Seasonings A
25g Pixian chili bean paste, 10g ground chilies, 6g fermented soy beans

Seasonings B
3g salt, 5g soy sauce, 1g MSG

Seasonings C
1g ground roasted Sichuan pepper, 200g everyday stock, 30g cornstarch-water mixture

Preparation
1. Cut the tofu into 1.8cm³ cubes, blanch in salty water, remove and soak in water.
2. Heat oil in a wok to 120℃, add Seasonings A and stir-fry to bring out the aroma. Add the stock, fried beef mince, season with Seasonings B, and simmer for 2 to 3 minutes; add the leeks and thicken with the cornstarch-water mixture; Transfer to a serving bowl and sprinkle with Seasonings C.

家常豆腐
Home-Style Tofu

色泽红亮，豆腐鲜嫩，味道浓香
pleasant reddish color; tender and soft tofu; savoury and aromatic taste

家常豆腐
Home-Style Tofu

[原 料]
豆腐300克 半肥瘦猪肉100克 青蒜苗50克 食用油1000克（约耗100克）

[调 料]
郫县豆瓣40克 食盐1克 味精3克 酱油10克 料酒15克 芝麻油5克 水淀粉10克 鲜汤200克

[制 作]
1. 豆腐切成长约6厘米、宽约3厘米、厚约0.6厘米的片；半肥瘦猪肉切成厚约0.2厘米的薄片；蒜苗斜切成节。
2. 锅置火上，下食用油烧至210℃，入豆腐片炸至色金黄时捞出。
3. 锅置火上，下食用油烧至160℃，入猪肉片炒香，续下郫县豆瓣炒香上色，掺入鲜汤，下豆腐、食盐、味精、酱油、料酒烧沸入味，最后下蒜苗，用水淀粉勾芡，淋入芝麻油即成。

Ingredients
300g tofu, 100g pork (with fat and lean meat), 50g baby leeks, 1000g cooking oil for deep-frying

Seasonings
40g Pixian chili bean paste, 1g salt, 3g MSG, 10g soy sauce, 15g Shaoxing cooking wine, 5g sesame oil, 10g cornstarch-water mixture, 200g everyday stock

Preparation
1. Cut the tofu into slices about 6cm long, 3cm wide and 0.6cm thick. Cut the pork into 0.2cm-thick slices and the leeks into sections.
2. Heat oil in a wok to 210℃, add tofu slices, fry till golen brown and remove.
3. Heat oil to 160℃, add pork slices and stir-fry to bring out the aroma. Add Pixian chili bean paste, stir well till aromatic. Add the stock, tofu, salt, MSG, soy sauce, Shaoxing cooking wine and bring to a boil. Wait till tofu absorbs the flavors, blend in the leeks, thicken with the cornstarch-water mixture and drizzle with the sesame oil.

口袋豆腐
Pocket Tofu

汤色乳白，豆腐形整不烂、内富浆汁，味浓而鲜

milky soup; juicy tofu strips; delicate and savoury taste

[原 料]

豆腐1000克 冬笋100克 菜心100克 食用油1500克（约耗70克） 奶汤1000克

[调 料]

食盐4克 胡椒粉1克 料酒15克 味精2克 鲜汤1000克 食用碱2克

[制 作]

1. 豆腐去外皮，切成长约6厘米、粗约2厘米见方的条；冬笋切片。
2. 锅中下食用油烧至210℃，入豆腐条炸至色金黄时捞出，放入加有食用碱的沸水里浸泡4~5分钟，然后捞入沸水锅中汆水2次，再用鲜汤汆制1次待用。
3. 锅置火上，掺奶汤烧沸，放入冬笋片、食盐、胡椒粉、料酒烧沸，下豆腐条、菜心、味精推匀，起锅装入汤盆中即成。

Ingredients

1000g tofu, 100g winter bamboo shoots, 100g tender leafy vegetables, 1500g cooking oil for deep-frying, 1000g milky stock

Seasonings

4g salt, 1g ground white pepper, 15g Shaoxing cooking wine, 2g MSG, 1000g everyday stock, 2g edible sodium carbonate

Preparation

1. Remove the surface skin of the tofu. Cut the tofu into strips about 6cm long, 2cm wide and 2cm thick. Slice the winter bamboo shoots.
2. Heat oil in a wok to 210℃ and deep-fry the tofu strips till golden brown. Transfer the tofu strips into boiling water with the edible sodium carbonate, and soak for 4 to 5 minutes. Remove the tofu strips, and then blanch them twice in water and once in stock.
3. Heat the milky stock in a wok, bring to a boil, and then add the bamboo shoot slices, salt, pepper and Shaoxing cooking wine. Bring to a second boil. Add the tofu, tender leafy vegetables and MSG, blend well and transfer to a soup bowl.

质地嫩滑，麻辣醇香
smooth and tender tofu; hot and spicy dipping sauce

过江豆花

Silken Tofu with Dipping Sauce

[原 料]
豆花500克

[调 料]
郫县豆瓣100克 豆豉蓉25克 辣椒粉50克 花椒粉10克 酱油50克 食盐4克 葱花30克 味精3克 芝麻油10克 食用油150克

[制 作]
1. 汤锅中掺清水，放入豆花煮沸后倒入大汤碗内。
2. 锅置火上，入食用油烧至120℃，下郫县豆瓣、豆豉蓉炒香上色，续下辣椒粉、花椒粉炒匀起锅装入碗中，加食盐、味精、酱油、芝麻油调匀成味汁，再装入小碟内，撒上葱花，与豆花一同上桌即成。

Ingredients
500g Silken Tofu

Seasonings
100g Pixian chili bean paste, 25g fermented soy beans (finely chopped), 50g ground chilies, 10g ground roasted Sichuan pepper, 50g soy sauce, 4g salt, 30g scallion (finely chopped), 3g MSG, 10g sesame oil, 150g cooking oil

Preparation
1. Heat the silken tofu in water, bring to a boil and then transfer to a large soup bowl.
2. Heat oil in a wok to 120℃, add the chili bean paste and fermented soy beans and stir-fry. Add the ground chilies, ground roasted Sichuan pepper, stir well and transfer to a bowl. Add the salt, MSG, soy sauce and sesame oil to the bowl, and stir well to make the dipping sauce. Transfer the dipping sauce onto small dipping saucers, and sprinkle with chopped scallion. Serve with the silken tofu.

用料丰富，营养均衡，汤菜合一，咸鲜醇厚
a variety of ingredients providing balanced nutrition; salty and savory taste

砂锅豆腐
Tofu Casserole

[原　料]

豆腐250克　熟鸡肉片30克　熟猪心片25克　熟猪肚片25克　水发金钩10克　火腿片25克　水发鱿鱼片60克　冬笋片20克　蘑菇片20克　鲜菜心50克　蕃茄片20克

[调　料]

食盐5克　味精5克　胡椒粉1克　姜片5克　葱段10克　奶汤750克　鸡油30克

[制　作]

1. 豆腐切成长约6厘米、宽约4厘米、厚约0.6厘米的片，入淡盐沸水中略煮后捞出备用。
2. 砂锅内放入姜片、葱段、冬笋片、蘑菇片、熟鸡肉片、熟猪心片、熟猪肚片、火腿片、金钩、豆腐，加入食盐、胡椒粉、味精、奶汤，置旺火上烧沸，再改用小火煮10分钟，之后加入鱿鱼片、鲜菜心、蕃茄片、鸡油煮2分钟即成。

Ingredients

250g tofu, 30g pre-cooked chicken (sliced), 25g pre-cooked pork heart (sliced), 25g pre-cooked pork tripe (sliced), 10g dried and shelled shrimps (soaked in water), 25g ham (sliced), 60g water-soaked squid (sliced), 20g winter bamboo shoots (sliced), 20g mushrooms (sliced), 50g tender leafy vegetables, 20g tomatoes (sliced)

Seasonings

5g salt, 5g MSG, 1g ground white pepper, 5g ginger (sliced), 10g scallion (cut into sections), 750g milky soup, 30g chicken oil

Preparation

1. Cut the tofu into slices about 6cm long, 4cm wide and 0.6cm thick, and then blanch in slightly salty water.
2. Add the ginger, scallion, bamboo shoots, mushrooms, chicken, pork heart, pork tripe, ham, shrimps, tofu, salt, ground white pepper, MSG, and milky soup to an earthen casserole. Bring to a boil and then simmer over a low flame for 10 minutes. Add the squid, vegetables, tomatoes and chicken oil, and then boil for two more minutes.

麻、辣、烫、鲜、香，风味别致
hot, spicy, pungent, aromatic, numbing and savory taste,

毛血旺
Duck Blood Curd in Chili Sauce

[原　料]

鸭血350克　黄豆芽300克　鳝鱼100克　午餐肉100克　猪心50克　火锅老油100克（食用油加辣椒、花椒、香料、姜、葱炒制而成）　食用油50克

[调料A]

干辣椒20克　花椒10克　火锅调料100克　葱段10克

[调料B]

食盐3克　料酒20克　味精2克　胡椒粉1克　芝麻油5克　鸡汤500克

[制　作]

1．鸭血切成厚片，入沸水锅中略煮至断生；其他原料均切成片。
2．锅中下食用油烧至150℃，放黄豆芽炒熟，装入大汤碗内垫底。
3．锅中下火锅老油烧至120℃，入调料A炒香，掺入鸡汤，续放调料B和各种原料，煮熟后倒入汤盆中即成。

Ingredients

350g duck blood curd, 300g soy bean sprouts, 100g paddy eels, 100g pork luncheon meat, 50g pork heart, 100g hot pot oil (stir-fry chili, Sichuan pepper, mixed herbal spice, ginger and scallion in cooking oil), 50g cooking oil

Seasonings A

20g dried chilies, 10g Sichuan pepper, 100g hot pot seasonings, 10g scallion (cut into sections)

Seasonings B

3g salt, 20g Shaoxing cooking wine, 2g MSG, 1g ground white pepper, 5g sesame oil, 500g chicken stock

Preparation

1. Cut the duck blood curd into chunks, boil briefly in a wok till just cooked and then slice the other ingredients.
2. Bring the cooking oil to 150℃, stir-fry the soy bean sprouts till cooked through and transfer to a large soup bowl.
3. Bring the hot pot oil to 120℃, add Seasonings A and stir-fry to bring out the aroma. Add Seasonings B and the remaining ingredients, boil till cooked through, and then transfer to the bowl.

汤汁清澈，咸鲜浓郁，菜嫩入味
crystal-clear soup; salty, delicate, savoury and refreshing taste; tender napa cabbage that has fully absorbed the flavors of the consommé

开水白菜
Napa Cabbage in Consomme

［原　料］

白菜心200克　清汤1500克

［调　料］

食盐2克　鸡精2克　胡椒粉1克　料酒10克

［制　作］

1. 将白菜心修切整齐，入沸水中略煮至断生时捞出，用冷水漂凉，整齐地放在汤碗内，加食盐、鸡精、胡椒粉、料酒、清汤，上笼用旺火蒸5分钟取出，沥去汤汁。
2. 锅置火上，入清汤烧沸，倒入盛菜心的碗内即成。

Ingredients

200g napa cabbage, 1500g consomme

Seasonings

2g salt, 2g chicken essence granules, 1g ground white pepper, 10g Shaoxing cooking wine

Preparation

1. Trim the cabbage, boil in water till just cooked, remove, rinse in cold water to cool, and then lay neatly in a soup bowl. Add the salt, chicken essence granules, ground white pepper, Shaoxing cooking wine and 250g consomme. Transfer to a steamer and steam for 5 minutes. Remove from the steamer and drain off the stock.
2. Heat the rest of the consomme in a wok and bring to a boil. Pour the boiling stock into the soup bowl containing the cabbage.

色泽红亮，外酥内嫩，味咸甜酸辣兼备，葱姜蒜香浓郁
brownish color; tender stuffing and crispy coating; strong aroma of ginger, scallion and garlic

鱼香茄饼

Eggplant Fritters in Fish-Flavor Sauce

[原　料]
茄子250克　猪肉250克　食用油1500克（约耗100克）

[调料A]
食盐1克　水淀粉15克　清水20克

[调料B]
全蛋淀粉糊250克　泡辣椒末40克　姜米10克　蒜米15克　葱花25克　食盐2克　酱油10克　白糖20克　醋25克　味精1克　水淀粉20克　鲜汤200克

[制　作]
1．猪肉剁碎，加入调料A拌匀成肉馅。
2．茄子去皮，切成两刀一断约1厘米厚的连夹片，将肉馅装入连夹片中。
3．锅中下食用油烧至150℃，将茄子连夹片裹上全蛋淀粉糊，放入油中炸至定型后捞出；待油温回升到180℃时，再将茄子放入炸至色金黄、质酥脆时捞出装入盘中。
4．锅中下食用油烧至120℃，放入泡辣椒末、姜米、蒜米、葱花炒香，续下鲜汤、食盐、酱油、白糖、醋、味精，用水淀粉收汁，浇淋在茄饼上即成。

Ingredients
250g eggplants, 250g pork, 1500g cooking oil for deep-frying

Seasonings A
1g salt, 15g cornstarch-water mixture, 20g water

Seasonings B
250g mixture of egg and cornstarch, 40g pickled chilies (finely chopped), 10g ginger (finely chopped), 15g garlic (finely chopped), 25g scallion (finely chopped), 2g salt, 10g soy sauce, 20g sugar, 25g vinegar, 1g MSG, 20g cornstarch-water mixture, 200g everyday stock

Preparation
1. Chop finely the pork and mix well with Seasonings A to make stuffing.
2. Peel the eggplants, cut into 2cm-thick sections, make one cut into each section and sandwich the stuffing in between.
3. Heat the oil in a wok to 150℃, coat the eggplant sections with the mixture of egg and cornstarch, and deep-fry them till the coating is cooked through. Remove from the oil. Reheat the oil to 180℃, deep-fry the fritters for a second time till their surfaces are golden brown and crispy, and then transfer to a serving dish.
4. Heat some oil in a wok to 120℃, add the pickled chilies, ginger, garlic and scallion, and stir-fry to bring out the aroma. Add the stock, salt, soy sauce, sugar, vinegar, MSG and the cornstarch-water mixture. Wait till the sauce thickens and pour over the eggplant fritters.

Steamed Egg with Topping
臊子蒸蛋

质地细嫩滑爽，肉末酥香爽口

tender and smooth egg; crispy and aromatic topping

[原 料]
猪肉末100克 鸡蛋180克

[调 料]
食盐2克 料酒10克 葱花5克 淀粉10克 猪油25克 清水100克

[制 作]
1. 锅置火上，入猪油烧至100℃，下猪肉末、料酒、食盐炒香成肉臊。
2. 鸡蛋破壳入碗，加清水、食盐、淀粉搅匀，倒入盘内，上笼蒸约6～7分钟后取出，浇上臊子，撒上葱花即成。

Ingredients
100g pork mince, 180g eggs

Seasonings
2g salt, 10g Shaoxing cooking wine, 5g scallion (finely chopped), 10g cornstarch, 25g lard, 100g water

Preparation
1. Heat lard in a wok to 100℃, slide in the pork mince, Shaoxing cooking wine, salt and stir-fry to make the topping.
2. Beat the eggs in a bowl, add water, salt and cornstarch, blend well and transfer to a plate. Put the plate in a steamer and steam for 6 to 7 minutes. Remove the plate from the steamer, pour the topping over the stamed egg and serve sprinkled with chopped scallion.

汁色乳白，咸鲜微苦，滋味清香
milky sauce; salty, delicate and slightly bitter taste; fragrant smell

白油苦笋

Stir-Fried Bitter Bamboo Shoots

[原 料]
鲜苦笋400克

[调 料]
食盐2克 味精1克 胡椒粉1克 料酒5克 水淀粉5克 奶汤150克 鸡油25克 芝麻油5克

[制 作]
1. 鲜苦笋去壳、剖开，入沸水锅中汆熟后用清水漂凉，再切成约0.2厘米厚的片。
2. 锅置火上，下奶汤、鸡油、食盐、胡椒粉、料酒、苦笋烧入味，再下味精、水淀粉、芝麻油，待汁液浓稠时起锅即成。

Ingredients
400g bitter bamboo shoots

Seasonings
2g salt, 1g MSG, 1g ground white pepper, 5g Shaoxing cooking wine, 5g cornstarch-water mixture, 150g milky stock, 25g chicken oil, 5g sesame oil

Preparation
1. Peel the bitter bamboo shoots, halve and blanch. Rinse them under cold water till cold and then cut into 0.2cm-thick slices.
2. Heat a wok over a flame, add the milky stock, chicken oil, salt, ground white pepper, Shaoxing cooking wine and bitter bamboo shoots, then braise till the bamboo shoots have absorbed the flavors of the condiments. Add the MSG, cornstarch-water mixture and sesame oil, and continue to simmer till the sauce thickens. Remove from the stove and transfer to a serving dish.

酱烧冬笋

Braised Winter Bamboo Shoots with Fermented Flour Paste

色泽棕红，酱香味浓
a rich red-brown color; the intense flavor of the fermented flour paste

酱烧冬笋

Braised Winter Bamboo Shoots with Fermented Flour Paste

[原 料]

冬笋500克 豌豆苗150克 化猪油1000克（约耗100克）

[调 料]

甜面酱50克 白糖5克 食盐2克 味精2克 芝麻油5克 鲜汤100克

[制 作]

1. 冬笋洗净，切成长约4厘米、粗约1厘米的条，放入160℃的化猪油中炸至表面皱皮后捞出。
2. 锅内留油少许，放入豌豆苗，加0.5克食盐翻炒断生，装入盘内垫底。
3. 锅内另下化猪油烧至100℃，放入甜酱炒香，加鲜汤，入冬笋、食盐、白糖烧至汁浓亮油，入味精、芝麻油炒匀，起锅装入盘内的豆苗上即成。

Ingredients

500g winter bamboo shoots, 150g pea vine sprouts, 1000g lard for deep-frying

Seasonings

50g fermented flour paste, 5g sugar, 2g salt, 2g MSG, 5g sesame oil, 100g everyday stock

Preparation

1. Rinse and cut the bamboo shoots into strips about 4cm long and 1cm thick. Deep-fry the strips in 160°C oil till their surfaces wrinkle. Remove.
2. Leave some oil in the wok, add pea vine sprouts and 0.5g salt, stir-fry till just cooked, and transfer to a serving dish.
3. Heat some oil in a wok to 100°C, add the fermented flour paste and stir-fry till fragrant. Add the stock, bamboo shoots, salt and sugar. Braise till the sauce is thick and lustrous, add the MSG and sesame oil, stir-fry to blend well and then tip over the pea vine sprouts on the dish.

色泽红亮，干香味浓
bright color; aromatic and savory taste

干锅茶树菇
Black Poplar Mushrooms in a Small Wok

典故

干锅是火锅演变出的一种形式，具备火锅的属性，只是在火锅的基础上减少了汤汁用量，突出原料本味，边煮、边炒、边食，汁愈来愈浓，渐入佳境。

Note

The type of dish developed from the hot pot and is referred to as a "dry-wok" (ganguo) dish in Chinese. Its difference from hot pot is that much less soup is added during the preparation and the liquid is gradually reduced so that little sauce remains when the dish is served.

[原　料]

干茶树菇100克　猪肉丝50克　食用油50克

[调　料]

食盐2克　味精1克　白糖1克　郫县豆瓣5克　胡椒粉1克　干辣椒段30克　葱段10克　酱油5克　辣椒油5克　芝麻油5克　鲜汤100克

[制　作]

1． 干茶树菇用温水泡发后去蒂，切成长约4厘米的段。
2． 锅置旺火上，下食用油烧热，入猪肉丝煸香、出油，续下郫县豆瓣、干辣椒段、葱段煸香，再下茶树菇煸干水分，加鲜汤、食盐、味精、白糖、胡椒粉、酱油，用小火烧制入味至汁干，最后淋入辣椒油和芝麻油，装入干锅内即成。

Ingredients

100g dried black poplar mushrooms, 50g pork slivers, 50g cooking oil

Seasonings

2g salt, 1g MSG, 1g sugar, 5g chili bean paste, 1g ground white pepper, 30g dried chilieses (cut into sections), 10g scallion (cut into sections), 5g soy sauce, 5g chili oil, 5g sesame oil, 100g stock

Preparation

1. Soak the dried black poplar mushrooms in warm water to reconstitute, remove the ends of their stems and then cut into 4cm lengths.
2. Heat oil in a wok over a high flame, toss in the pork slivers and stir-fry till aromatic. Add the chili bean paste, dried chilieses, scallion and stir-fry till aromatic. Blend in the mushrooms and stir-fry till they have lost their water content. Add the stock, salt, MSG, sugar, ground white pepper and soy sauce, and simmer till the sauce has been absorbed. Drizzle with chili oil and sesame oil and transfer to a small wok for serving.

干锅茶树菇
Black Poplar Mushrooms in a Small Wok

咸鲜清香，清淡爽口
salty, delicate and aromatic taste

干贝菜心
Napa Cabbage with Dried Scallops

[原　料]

白菜心600克　干贝25克　化猪油75克

[调　料]

姜片15克　葱段20克　食盐3克　料酒5克　味精1克　奶汤250克　水淀粉25克　化鸡油15克

[制　作]

1. 干贝用热水洗净，上笼蒸软待用；白菜心洗净、去筋，入沸水锅内汆至断生，再放入清水中漂冷后捞起沥干水分。
2. 炒锅置旺火上，下化猪油烧至120℃，放入姜片、葱段炒香，掺奶汤烧沸，捞出姜片、葱段，放入菜心、干贝（连汁）、料酒、食盐烧入味，再入味精，用水淀粉勾薄芡，淋上化鸡油，起锅装盘即成。

Ingredients

600g napa cabbage, 25g dried scallops, 75g lard

Seasonings

15g ginger (sliced), 20g scallion (cut into sections), 3g salt, 5g Shaoxing cooking wine, 1g MSG, 250g milky stock, 25g cornstarch-water mixture, 15g chicken oil

Preparation

1. Rinse the dried scallops with hot water, and steam till soft. Wash the cabbage, remove any strings and blanch till al dente. Transfer the cabbage into cold water to cool. Remove and drain.
2. Heat lard in a wok to 120°C, add the ginger and scallion, and then stir-fry till aromatic. Add the milky stock and bring to a boil. Remove and discard the ginger and scallion, and add the cabbage, scallops, Shaoxing cooking wine and salt to braise till the vegetables have absorbed the flavors of the sauce. Pour in the cornstarch-water mixture to thicken the sauce. Drizzle with chicken oil and transfer to a serving dish.

干煸四季豆
Dry-Fried French Beans

色泽碧绿，干香细嫩，咸鲜香浓
verdant color; crispy but tender beans salty, savoury and aromatic taste

[原 料]
四季豆300克 猪肉臊50克 芽菜末15克 食用油1000克(约耗50克)

[调 料]
食盐2克 味精1克 芝麻油3克

[制 作]
1. 四季豆去筋洗净，切成长约8厘米的节。
2. 锅中下食用油烧至150℃，放入四季豆炸至断生、皱皮，呈油绿色时捞出。
3. 锅中留油少许，放入四季豆、猪肉臊、芽菜末、食盐、味精炒出香味，淋入芝麻油，起锅装盘即成。

Ingredients
300g French beans, 50g stir-fried pork mince, 15g yacai (preserved mustard stems minced), 1000g cooking oil for deep-frying

Seasonings
2g salt, 1g MSG, 3g sesame oil

Preparation
1. Remove the strings of the French beans, wash and chop into 8cm lengths.
2. Heat the oil in a wok to 150℃, add the French beans, deep-fry till just cooked and wrinkled, and then remove.
3. Leave some oil in the wok, return the French beans to the wok, add the stir-fried pork mince, yacai, salt and MSG to stir-fry till aromatic. Drizzle with sesame oil and transfer to a serving dish.

色泽淡雅，甜香不腻
fragrant and delicate flavor; sweet and mild taste

蚕豆泥
Mashed Broad Beans

[原 料]
蚕豆500克 化猪油150克

[调 料]
白糖50克

[制 作]
1. 蚕豆入锅煮熟，捞出沥干水分，用搅拌机搅成泥蓉状。
2. 锅置火上，下化猪油、蚕豆泥反复翻炒至翻砂，放入白糖炒匀装盘即成。

Ingredients
500g broad beans, 150g lard

Seasoning
50g sugar

Preparation
1. Boil the broad beans till fully cooked. Remove, drain and then mash into a paste in a food-processor.
2. Heat a wok over a flame, add oil, and stir-fry the broad bean paste till its water content has evaporated and it is no longer sticky. Add sugar, blend well and transfer to a plate.

汤汁香甜，雪梨滋糯
fragrant and luscious juice; glutinous and smooth pears

川贝酿雪梨
Pear Stuffed with Fritillaria Cirrhosa

[原 料]
雪梨4个　川贝母6克　糯米30克　苡仁20克　蜜饯瓜条20克

[调 料]
冰糖细粒50克

[制 作]
1. 糯米、苡仁洗净，用清水泡涨后捞出，沥干水分。
2. 雪梨去皮，在三分之一处切下梨蒂作盖，用小勺挖去梨核，浸泡在清水中，再入沸水中煮1分钟捞出，用冷水冲凉。
3. 蜜饯瓜条切成小颗粒，与糯米、苡仁、冰糖拌匀后装入梨内，再放入川贝母，盖上梨盖，放入蒸碗，用湿绵纸封住碗口，入笼用旺火蒸1小时后出笼。
4. 将蒸梨的原汁倒入锅中，加清水少许，放入冰糖熬化收稠，浇淋在梨上即成。

Ingredients
4 pears, 6g Fritillaria cirrhosa, 30g glutinous rice, 20g coix seeds, 20g candied melon strips

Seasonings
50g rock sugar (crushed)

Peparation
1. Wash the glutinous rice and coix seeds, soak in water till swollen, remove and drain.
2. Peel the pears, and cut off the top third of each (with the stems) to serve as covers. Core with a spoon and soak the pears in water. Transfer the pears into boiling water, boil for 1 minute and then rinse with cold water till cool.
3. Chop the candied ash gourd strips, mix with the glutinous rice, coix seeds and rock sugar, and blend well. Transfer the mixture into the pears. Put pieces of Fritillaria cirrhosa onto the stuffings and cover. Lay the pears in a steaming bowl, seal the bowl with wet cotton paper towels. Steam over a high flame for one hour and then set aside.
4. Pour the water left in the steaming bowl into a wok, add a little water and some rock sugar, then simmer till the sugar melts and the soup becomes viscous and thick. Pour the soup over the pears.

川贝酿雪梨
Pear Stuffed with Fritillaria Cirrhosa

色泽金黄，酥香咸鲜
golden brown color; aromatic and mild taste; yolk melting in the mouth

金沙玉米

Golden-Sand Corn (Fried Corn with Egg Yolk)

［原　料］

罐头玉米粒250克　食用油500克（约耗30克）

［调　料］

咸鸭蛋黄4枚　食盐0.5克　干细淀粉25克

［制　作］

1. 玉米粒洗净后沥干水分，拌上淀粉，入150℃的食用油中炸至色金黄、质酥脆时捞出；咸鸭蛋黄捣烂成泥。
2. 锅置旺火上，入少许食用油烧至120℃，下咸鸭蛋黄泥炒翻砂，续下食盐、玉米粒，用小火翻炒至玉米粒粘裹上咸鸭蛋黄并呈微黄色时起锅装盘即成。

Ingredients

250g canned corn, 500g oil for deep-frying

Seasonings

4 salted duck egg yolks (mashed), 0.5g salt, 25g cornstarch

Preparation

1. Drain the corn, mix with the cornstarch and fry in 150℃ oil till golden brown and crisp. Remove; Mash the yolk.
2. Heat some oil in a wok over a high flame to 120℃, add the mashed egg yolks and stir-fry till the yolk is no longer sticky and resembles the texture of sand. Add the salt and corn, and continue to stir-fry over a low flame till the corn is coated with the yolk and becomes golden. Remove from the heat and transfer to a serving dish.

红白绿相间，咸鲜清淡
contrasting colors of red, white and green; salty and delicate taste

番茄蛋花汤
Tomato and Egg Soup

[原 料]
鸡蛋2个 番茄200克 小白菜50克 鲜汤750克 食用油10克

[调 料]
食盐3克 味精1克 水淀粉40克

[制 作]
1. 番茄洗净后切成丁；小白菜洗净后切成碎片；鸡蛋入碗搅散。
2. 锅中放鲜汤烧沸，入食盐、味精、番茄、小白菜、食用油烧沸，放入水淀粉搅匀，再放入鸡蛋液推匀，起锅装入汤碗中即成。

Ingredients
2 eggs, 200g tomatoes, 50g bok choy, 750g everyday stock, 10g cooking oil

Seasonings
3g salt, 1g MSG, 40g cornstarch-water mixture

Preparation
1. Wash the tomatoes thoroughly and dice. Wash the bok choy thoroughly and chop into small pieces. Beat the eggs in a bowl.
2. Bring the stock to a boil in a pot, add the salt, MSG, tomatoes, bok choy and cooking oil, and bring to a second boil. Add the cornstarch-water mixture and blend well. Pour in the beaten egg, and stir well. Transfer to a serving bowl.

绿豆南瓜汤

Pumpkin Soup with Mung Beans

清甜润滑，本味突出
pure, delicate and natural taste

绿豆南瓜汤
Pumpkin Soup with Mung Beans

[原　料]
干绿豆60克　老南瓜1000克

[调　料]
白糖60克

[制　作]
1. 将干绿豆用清水浸泡至吸满水分后捞出待用；老南瓜去皮及瓜瓤，切成约3厘米见方的块。
2. 砂锅内放入清水烧沸，下绿豆，加盖用小火煮30分钟，再放入南瓜煮30分钟，至绿豆开花、南瓜软熟时加少许白糖调味即成。

Ingredients
60g dried mung beans, 1000g pumpkin

Seasonings
60g sugar

Preparation
1. Soak the mung beans in water, wait till the beans have absorbed the water, and drain. Peel the pumpkin, remove the seeds, and then cut into 3cm³ cubes.
2. Bring water to a boil in an earthen pot, add the mung beans, and simmer over a low flame for 30 minutes. Add the pumpkin and boil for another 30 minutes till the mung beans and the pumpkin cubes are cooked through. Add the sugar.

Sichuan (China) Cuisine in Both Chinese and English

川菜
（中英文标准对照版）

火 锅
Hot Pot

第三篇

用料、质感多样，自烫自食，麻辣鲜香，风味别致
diversified ingredients; pungent, aromatic and appetizing taste

毛肚火锅
Beef Tripe Hot Pot

[原 料]
毛肚、鸭肠、鸭胗、鸭血、嫩牛肉、猪腰、猪脑、鳝鱼、鱼肉、菇类、菌类、豆制品、粉皮、蔬菜等适量

[调料A]
郫县豆瓣120克 豆豉20克 辣椒粉25克 花椒粉10克 干辣椒40克 花椒15克 生姜30克 葱段30克 芽菜30克 香料40克 食盐5克 料酒40克 白糖10克 牛油600克 熟菜油400克 鲜汤1500克

[调料B]
芝麻油20克 食盐1克 味精1克 蒜泥10克

[制 作]
1. 锅中放熟菜油烧至120℃，放入郫县豆瓣、香料、生姜、葱段炒香，再放入豆豉、芽菜、辣椒粉炒香，掺入鲜汤烧沸，下食盐、料酒、白糖、花椒粉、味精调制成麻辣味的卤汁。
2. 牛油入锅烧化，放入干辣椒、花椒炒香，倒入火锅卤汁熬制出味。
3. 将调料B放入碟中拌匀成味汁，与原料一同上桌即成。

Ingredients
cattle tripe, duck intestines, duck gizzards, duck blood curd, tender beef, pork kidneys, pork brain, paddy eel, fish, mushrooms, funguses, various kinds of toufu, steamed mung been jelly, vegetables

Seasonings A
120 Pixian chili bean paste, 20g fermented soy beans, 25g ground chilies, 10g ground roasted Sichuan pepper, 40g dried chilies, 15g Sichuan pepper, 30g ginger, 30g scallion (cut into sections), 30g yacai (preserved mustard stems minced), 40g mixed herbal spices, 5g salt, 40g Shaoxing cooking wine, 10g sugar, 600g beef dripping, 400g precooked rapeseeds oil, 1500g everyday stock

Seasonings B
20g sesame oil, 1g salt, 1g MSG, 10g garlic (finely chopped)

Preparation
1. Heat oil in a wok to 120℃, and stir-fry chili bean paste, herbal spices, ginger and scallion till aromatic. Add fermented soy beans, yacai and ground chilies to stir-fry. Pour in the stock, bring to a boil, and add salt, Shaoxing cooking wine, sugar, ground roasted Sichuan pepper and MSG to make hot-and-spicy broth.
2. Melt the beef dripping in a wok, stir-fry dried chilies and Sichuan pepper till aromatic, add the hot-and-spicy broth and simmer to bring out the fragrance.
3. Mix Seasonings B, and blend well to make the dipping sauce.

气氛热烈，口感丰富，咸鲜麻辣烫
wide range of selection; delicate milky soup and pugent spicy soup

鸳鸯火锅

Double-Flavor Hot Pot

[原　料]
水发毛肚、鸭肠、鳝鱼片、嫩牛肉、鱼肉、猪腰片、鲜虾、鸭血、蔬菜等适量

[红汤火锅汤料]
毛肚火锅汤料1500克　干辣椒20克　花椒10克

[白汤火锅汤料]
高汤1500克（用猪棒骨和鸡一起熬制出来的汤）　生姜10克　葱段20克　番茄片40克　食盐10克　料酒20克　鸡精2克　胡椒粉3克　鸡化油20克

[制　作]
1. 各种食用原料经刀工处理后分别装盘。
2. 将红汤火锅汤料和白汤火锅汤料分别放入鸳鸯火锅的两个"S"格中，上桌点火，待火锅汤汁沸腾后烫食即可。

Ingredients
water-soaked cattle tripe, duck intestines, paddy eel (sliced), tender beef, fish, pork kidneys (sliced), prawns, duck blood curd, vegetables

Spicy Soup
1500g hot-and-spicy broth of Sichuan style, 20g dried chilies, 10g Sichuan pepper

Milky Soup
1500g stock (made by simmering pork leg bones and chicken), 10g ginger, 20g scallion (cut into sections), 40g tomatoes (sliced), 10g salt, 20g Shaoxing cooking wine, 2g chicken essence granules, 3g ground white pepper, 20g chicken oil

Preparation
1. Cut the ingredients and transfer to plates.
2. Add the two kinds of soup respectively to the two different compartments of a pot partitioned by an "S"-shaped dividing bar. Bring to a boil and boil the ingredients in the soup.

汤色乳白，肉质细嫩，肥而不腻
milky soup; tender, fatty but not greasy meat

羊肉汤锅
Mutton Soup Hot Pot

[原 料]
羊腿肉2500克 羊杂1000克 羊头1个 羊骨5000克 萝卜片100克 白菜100克 豌豆苗100克 生菜100克

[调 料]
食盐10克 胡椒粉3克 姜片50克 葱段60克 白酒20克 味精10克 料酒20克 枸杞10克 红枣10克

[味碟料]
香葱花60克 香菜60克 红尖椒40克 豆腐乳60克 食盐6克 味精6克

[制 作]
1. 分别将羊腿肉、羊杂、羊头、羊骨入锅氽水后备用。
2. 大汤锅置火上，放入清水、羊头、羊骨、羊腿肉、羊杂、姜片、葱段、白酒，用旺火烧沸，去掉浮沫，改用中火煮至羊肉、羊杂软熟时捞出，晾凉改刀，锅中羊骨再煮3～5小时至汤色乳白、香味浓郁。
3. 味碟料分别装入6个小碗中作味碟。
4. 取一汤锅，掺入羊肉汤，放入食盐、胡椒粉、姜片、葱段、味精、料酒、枸杞、红枣调成汤汁，上桌后点火烧沸，配熟羊肉片、羊杂和四个时鲜蔬菜，烫涮原料后蘸味碟食用。

Ingredients
2500g mutton leg meat, 100g mutton offal, 1 mutton head, 5000g mutton bones, four kinds of vegetables (100g radish, 100g celery cabbage, 100g pea vine spourts, 100g lettuce)

Seasonings
10g salt, 3g ground white pepper, 50g ginger (sliced), 60g scallion (cut into sections), 20g alcohol, 10g MSG, 20g Shaoxing cooking wine, 10g wolfberries, 10g red dates

Dipping Sauce
60g scallion (chopped), 60g coriander leaves, 40g red chili peppers, 60g fermented tofu, 6g salt, 6g MSG

Preparation
1. Blanch the mutton leg meat, mutton offal, mutton bones and mutton head.
2. Put a big soup pot over a flame, add water, mutton head, mutton bones, mutton leg meat, mutton offal, ginger, scallion and alcolhol, bring to a boil, remove the scums and simmer over a medium flame till the mutton and offal are soft and cooked through. Remove the mutton and offal, and slice them. Boil the bones for another three to five hours till the soup becomes milky and aromatic.
3. Divide the dipping sauce into six portions and transfer to six small bowls.
4. Add the milky soup, salt, pepper, gingr, scallion, MSG, Shaoxing cooking wine, wolfberries and dates to a pot. Serve the pot, heat over a flame and bring to a boil. Boil the cooked mutton slices, offal slices and vegetables, and dip into the the dipping sauce before eating.

串串香
Chuan Chuan Xiang Hot Pot

麻、辣、烫，味道浓厚，用料多样，自烫自食
spicy, pungent and hot taste; varied ingredients

串串香

Chuan Chuan Xiang Hot Pot

[原　料]

毛肚、鸭肠、鸡翅、牛肉、兔腰、鳝鱼、鹌鹑蛋、豆制品、蔬菜等适量

[调料A]

郫县豆瓣150克　豆豉15克　辣椒粉150克　干辣椒30克　花椒20克　生姜30克　葱段30克　香料20克　食盐10克　味精5克　料酒50克　冰糖10克　醪糟汁30克　牛油200克　熟菜子油400克　鲜汤1500克

[调料B]

食盐2克　味精1克　辣椒粉15克　花椒粉2克　熟芝麻10克　酥花生碎米6克

[制　作]

1. 锅中下牛油烧化，放干辣椒节炒至色棕红时放花椒炒香，连油带料倒入大碗中成牛油汤汁。
2. 锅中放熟菜子油烧至120℃，放入调料A炒香，加入鲜汤和牛油汤汁，用中小火熬制到香味浓郁时加入味精成卤汁。食用时将原料用竹签穿上，放在卤汁中涮熟后捞出，拌上调料B食用即可。

典故

串串香最早出现在20世纪80年代中期的四川成都。该菜是将原料用竹签串上，放入调制好的麻辣味汤卤中烫熟后食用，因此得名。

Note

Chuan Chuan Xiang hot pot originated from the mid 1980s in Chengdu. The ingredients are first skewered with bamboo sticks (known as "Chuan Chuan" in Chinese), then cooked in the hot-and-spicy broth. The dish is so called because each Chuan Chuan is aromatic and savoury ("Xiang" in Chinese).

Ingredients

beef tripe, duck intestines, chicken wings, beef, rabbit kidneys, paddy eel, quail eggs, various kinds of tofu, various vegetables

Seasonings A

150g Pixian chili bean paste, 15g fermented soy beans, 150g ground chilies, 30g dried chilieses, 20g Sichuan pepper, 30g ginger, 30g scallion (cut into sections), 20g mixed herbal spices, 10g salt, 5g MSG, 50g Shaoxing cooking wine, 10g rock sugar, 30g fermented glutinous rice wine, 200g beef dripping, 400g precooked rapeseed oil, 1500g everyday stock

Seasonings B

2g salt, 1g MSG, 15g ground chilies, 2g ground roasted Sichuan pepper, 10g roasted sesame seeds, 6g crispy peanuts (roasted or fried finely chopped)

Preparation

1. Melt the beef dripping in a wok over a flame, stir-fry the dried chilies till dark brown, then add the Sichuan pepper and stir-fry till aromatic. Transfer this sauce into a large bowl.
2. Heat oil in a wok to 120℃, add Seasonings A and stir-fry till aromatic. Add the stock and prepared sauce, and simmer over a medium-low flame till aromatic. Add MSG to finish the broth. Skewer the ingredients with bamboo sticks, boil in the broth till cooked through and then mix well with Seasonings B.

色泽红亮，鱼肉细嫩，青花椒味浓厚
tingling taste of the green Sichuan pepper; fiery and bright color; tender and aromatic fish

冷锅鱼
Fish in Cold Pot

典故
冷锅是火锅与干锅的结合，是将食物烹制好后连锅一起端上桌，不点火加热而供客人食用；待客人将锅中食物吃完之后再点火加热，以烫食其他食物。

Note
Cold pot is here used in contrast to the term hot pot. Cold pot refers to the serving of cooked food in a pot, which is ready to be eaten without having to be boiled like hot pot. When the ingredients in the pot have been eaten, the pot can then be heated to boil other ingredients to your liking.

[原　料]
花鲢鱼片1500克　牛骨汤2000克

[调料A]
食用油400克　红油豆瓣200克　泡红辣椒400克　泡仔姜30克　榨菜50克　干青花椒40克　八角1个　葱段20克　芝麻油10克　香料粉3克

[调料B]
酥黄豆10克　香菜6克　大头菜10克　榨菜5克　蒜泥5克　香葱10克

[制　作]
1. 锅中下食用油烧至100℃，放入红油豆瓣、泡红辣椒、干青花椒炒10分钟，再放入八角炒20分钟，加入榨菜、泡仔姜、葱段炒约2分钟出香后掺入牛骨汤，烧5分钟至出香味后沥去料渣，放入香料粉、芝麻油、鱼片，待鱼片煮至刚熟时倒入火锅中上桌。
2. 将调料B混合后放入调味碗中，食用时加一勺锅里的原汁，一人一碟。

Ingredients
1500g silver carp (sliced), 2000 beef bone soup

Seasonings A
400g cooking oil, 200g chili bean paste, 400g pickled chilies, 30g pickled tender ginger, 50g zhacai (preserved mustard tuber sliced), 40g dried green Sichuan pepper, 1 star anise, 20g scallion (cut into sections), 10g sesame oil, 3g ground mixed herbal spices

Seasonings B
10g roasted soy bean, 6g coriander leaves, 10g preserved kohlrabi, 5g zhacai (preserved mustard tuber, finely chopped), 5g garlic (finely chopped), 10g scallion

Preparation
1. Heat oil in a wok to 100℃, add chili bean paste, pickled chilies and dried green Sichuan pepper, and stir-fry for 10 minutes. Add the star anise, and stir-fry for 20 minutes. Add zhacai, pickled ginger and scallion, then stir-fry for 2 minutes to bring out the aroma. Add the beef bone soup, and simmer for 5 minutes. Remove the scums and add the herbal spices, sesame oil and fish slices. Boil till the fish is just cooked. Remove from the stove and pour the contents in the wok into a serving pot.
2. Mix Seasonings B in a bowl. Add the soup in the serving pot to the mixture when eating.

冷锅鱼 Fish in Cold Pot

色彩鲜艳，肉质干香，咸鲜麻辣

bright and lustrous color; aromatic chicken; salty, numbing and pungent taste

干锅鸡

Sauteed Chicken in a Small Wok

[原 料]
仔公鸡1000克 青尖椒、红尖椒各50克 西芹40克 洋葱40克 酥花仁50克 食用油1000克（约耗75克）

[调料A]
食盐1克 姜片6克 葱段10克 料酒10克

[调料B]
干辣椒段30克 花椒10克 姜片10克 蒜片10克 葱丁15克 郫县豆瓣20克 食盐3克 酱油5克 白糖2克 味精1克 料酒15克 鲜汤100克

[制 作]
1. 仔公鸡斩成约2.5厘米见方的块，用调料A拌匀码味15分钟；青尖椒、红尖椒、西芹、洋葱均切成长约2厘米的段。
2. 锅置旺火上，下食用油烧至180℃，入鸡块炸干水分且呈棕红色时捞出。
3. 锅置旺火上，下食用油烧至120℃，放入干辣椒、花椒、郫县豆瓣、姜片、蒜片炒香，续下鸡块、青尖椒、红尖椒、洋葱、西芹，掺鲜汤，再下食盐、料酒、白糖、酱油、葱丁、味精、酥花仁炒匀，装入锅仔中即成。

Ingredients
100g poussin rooster, 50g red chili peppers, 50g green chili peppers, 40g celery, 40g onions, 50g crispy peanuts (fired or roasted), 1000g cooking oil for deep-frying, 100g everyday stock

Seasonings A
1g salt, 6g ginger, sliced, 10g scallion (cut into sections), 10g Shaoxing cooking wine

Seasonings B
30g dried chiliesies (cut into sections), 10g Sichuan pepper, 10g ginger (sliced), 10g garlic (sliced), 15g scallion (finely chopped), 20g Pixian chili bean paste, 3g salt, 5g soy sauce, 2g sugar, 1g MSG, 15g Shaoxing cooking wine

Preparation
1. Cut the chicken into 2.5cm³ cubes, blend with Seasonings A and marinate for 15 minutes. Chop the red chili peppers, green chili peppers, celery and onions into sections about 2cm long.
2. Heat a wok over a high flame, add the oil and go on heating the oil to 180℃. Deep-fry the chicken to evaporate the water content till the chicken cubes become brown. Remove the chicken from the oil.
3. Heat some oil in a wok over a high flame to 120℃, add the dried chilies, Sichuan pepper, chili bean paste, ginger and garlic, and then stir-fry to bring out the aroma. Add the chicken cubes, red chili peppers, green chili peppers, onions, celery, stock, salt, Shaoxing cooking wine, sugar, soy sauce, scallion, MSG and peanuts, and stir to mix them evenly. Transfer to a small wok for serving.

Sichuan (China) Cuisine in Both Chinese and English

川菜
(中英文标准对照版)

面点小吃
Snacks

第四篇

面臊酥香，面条滑爽，咸鲜微辣，芽菜香浓
crispy pork mince; smooth and springy noodles; salty, savoury and slightly hot taste; strong aroma from yacai

担担面
Dandan Noodles

[原　料]

面条500克　半肥瘦猪肉粒200克　猪油25克

[调　料]

酱油50克　食盐1克　红油辣椒25克　芽菜25克　味精15克　葱花50克　料酒15克　食醋15克　甜酱10克　鲜汤200克　白糖5克

[制　作]

1. 锅置中火上，入猪油烧至150℃，下猪肉粒炒散籽，加料酒、甜酱、食盐炒至猪肉吐油、香酥时起锅成面臊。
2. 将食盐、酱油、味精、红油辣椒、芽菜、葱花、食醋、白糖加鲜汤调为滋汁。
3. 锅置火上，入水烧沸，下面条煮熟，捞出后盛入调好味的碗中，舀入面臊即成。

Ingredients

500g noodles, 200g pork mince (a mixture of fat and lean meat), 25g lard

Seasonings

50g soy sauce, 1g salt, 25g oil-infused chili flakes, 25g yacai (preserved mustard stems, minced), 15g MSG, 50g scallion (finely chopped), 15g Shaoxing cooking wine, 15g vinegar, 10g fermented flour paste, 200g everyday stock

Preparation

1. Heat the lard in a wok over a medium flame till 150℃, slide in the pork mince and stir-fry to separate. Add the Shaoxing cooking wine, fermented flour paste and salt, and continue to stir-fry till the pork becomes crispy and aromatic. The crispy pork mince will serve as the topping for the noodles.
2. Mix the salt, soy sauce, MSG, oil-infused chili flakes, yacai, scallion, vinegar and sugar in a serving bowl, then add the stock.
3. Heat some water in a wok, bring to a boil, add the noodles, and continue to boil till the noodles are cooked through. Transfer the noodles into the serving bowl and top with the stir-fried pork mince.

钟水饺
Zhong's Dumplings

色泽美观，柔韧爽滑，麻辣香浓

bright and lustrous color; smooth and chewy dumplings; moreish and pungent taste

[原料]

面粉250克 猪后腿肉末250克 鸡蛋1个

[调料]

生姜25克 葱25克 复制酱油100克 红油辣椒75克 食盐2克 料酒25克 味精2克 蒜泥50克 胡椒粉1克 芝麻油5克 熟芝麻5克

[制作]

1. 面粉加清水调揉成面团，并盖上湿毛巾饧面。
2. 生姜拍破，葱挽结，用清水浸泡成姜葱水；猪肉置盆内，加少许姜葱水、料酒、味精、胡椒粉、食盐、鸡蛋、芝麻油、鲜汤用力搅匀成粘稠状的馅心。
3. 将面团搓成圆条，再扯成剂子，擀成直径为5厘米的圆形皮坯，放入肉馅，捏成半月形成生饺。
4. 用旺火沸水煮饺，待饺皮起皱发亮即熟；用漏瓢捞出熟饺盛入碗内，加入复制酱油、味精、红油辣椒、蒜泥、熟芝麻即成。

Ingredients

250g flour, 250g pork hind leg meat (minced), 1 egg

Seasonings

25g ginger, 25g scallion, 100g concocted soy sauce, 75g oil-infused chili flakes, 2g salt, 2g MSG, 25g Shaoxing cooking wine, 50g garlic (finely chopped), 1g pepper, 5g sesame oil, 5g roasted sesame seeds

Preparation

1. Mix the flour and water to make a dough. Cover the dough with a towel and let stand for a while.
2. Smash the ginger, tie the scallion into knots and soak them in water to make ginger-and-scallion-flavored water. Mix the pork, some ginger-and-scallion-flavored water, Shaoxing cooking wine, MSG, pepper, salt, egg, sesame oil and stock to make sticky stuffing.
3. Knead the dough into long rolls, tear into small portions and then flatten them with a rolling pin into thin round wrappers about 5cm in diameter. Put some stuffing onto a wrapper, and then fold the wrapper to wrap the stuffing up (shaped like a first quarter moon). Press hard to secure the edges.
4. Boil the dumplings over a high flame till they float and their wrappers become wrinkled and transparent. Transfer the dumplings with a perforated ladle to a bowl, and add the concocted soy sauce, MSG, oil-infused chili flakes, garlic and sesame seeds.

色泽美观,柔韧爽滑,麻辣香浓
pleasant color; smooth and springy jelly; spicy and zingy taste

川北凉粉
Northern-Sichuan-Style Pea Jelly

[原 料]

凉粉500克

[调 料]

食盐5克 味精2克 辣椒油75克 花椒粉5克 蒜泥25克 豆豉蓉50克 水淀粉25克 熟菜油20克 葱花10克

[制 作]

1. 锅置火上,放熟菜油烧热,下豆豉蓉、食盐炒香,加水煮5~6分钟,再用水淀粉和味精勾成二流芡,起锅、冷却后成豆豉卤。
2. 将凉粉切成长条放入盘中,淋上豆豉卤、花椒粉、辣椒油、葱花和蒜泥即成。

Ingredients

500g pea jelly

Seasonings

5g salt, 2g MSG, 75g chili oil, 5g ground roasted Sichuan pepper, 25g garlic (finely chopped), 50g fermented soy beans (finely chopped), 25g cornstarch-water mixture, 20g pre-cooked rapeseed oil, 10g scallion (chopped)

Preparation

1. Heat the oil in a wok, add the fermented soy beans and salt, and then stir-fry till aromatic. Add some water, boil for 5 to 6 minutes, and then blend in the cornstarch-water mixture and MSG. Stir well, remove from the heat and cool. The sauce made this way is called fermented-soy-bean-flavored sauce.
2. Cut the pea jelly first into thin slices and then into strips, transfer to a serving dish, pour over the fermented-soy-bean-flavored sauce, and drizzle with ground roasted Sichuan pepper, chili oil, scallion and chopped garlic.

皮薄滑爽，馅嫩化渣，汤色奶白，香味醇厚
thin and smooth wrappers; juicy and tender stuffing; milky soup and aromatic taste

龙抄手
Long Wonton

典故

成都名小吃。创办者于浓花茶园商议开店事宜，在议招牌时，借用了浓花的"浓"之谐音"龙"冠于抄手之前，故名"龙抄手"。

Note

The dish is a famous Chengdu snack. Before the restaurant was opened, the founders used to discuss relevant issues in a tea house named Nong Hua. They later used the Chinese character "龙"(referring to dragon in Chinese), a word whose pronunciation is quite similar to Nong, to distinguish their wonton from others.

[原 料]

手工抄手皮100张 猪肉泥500克 鸡蛋1个 原汤2000克

[调 料]

生姜10克 胡椒粉2克 芝麻油15克 味精5克 食盐10克 料酒15克

[制 作]

1. 猪肉泥入碗，加食盐、生姜水搅匀，再加鸡蛋液、胡椒粉、料酒、芝麻油、味精继续搅拌至粘稠状成馅心，放在抄手皮上，包折成抄手生坯。
2. 食盐、胡椒、味精和适量原汤装入碗内。
3. 锅置旺火上，入水烧沸，下抄手生坯煮熟，捞出后盛入调好味的碗中即成。

Ingredients

100 hand-made wonton wrappers, 500g pork mince, 1 egg, 2000g stock

Seasonings

10g ginger, 2g ground white pepper, 15g sesame oil, 5g MSG, 10g salt, 15g Shaoxing cooking wine

Preparation

1. Lay the pork mince into a bowl, add salt and ginger-soaked water, and then blend well. Add then beaten egg, ground white pepper, Shaoxing cooking wine, sesame oil and MSG, then whisk till the mixture becomes sticky paste, serving as the stuffing. Put some stuffing onto a wrapper, then fold and squeeze to seal the wrapper to make wonton.
2. Add salt, pepper, MSG and some stock to a bowl.
3. Heat water in a wok over a high flame, bring to a boil and dump the wontons in. Continue to boil till the wontons are cooked through, remove and transfer to the bowl.

Crispy Pancakes with Beef Stuffing

牛肉焦饼

皮酥脆爽口，味咸鲜微麻，馅细嫩渣
flaky and crunchy crust; tender stuffing; salty, aromatic and numbing taste

牛肉焦饼
Crispy Pancakes with Beef Stuffing

[原　料]
面粉500克　黄牛腿肉400克　菜子油1500克

[调　料]
食盐10克　葱300克　味精2克　生姜7克　花椒10克　醪糟汁15克

[制　作]
1. 生姜与花椒混合剁成细姜椒末；葱切花；牛腿肉洗净去筋，剁成细颗粒，加入食盐拌匀，再加菜子油200克、醪糟汁、姜椒末、味精、葱花拌匀成馅心。
2. 面粉用清水调制成团，盖上湿毛巾饧面30分钟，然后搓成圆条，切成剂子，擀成约1毫米厚、40厘米长的薄片成皮坯。
3. 取皮坯一张，涂抹上少许菜子油，将馅心放于皮坯的一端，从有馅的一端起，将皮坯卷成圆筒，边卷边把卷筒两端稍作延伸，卷成筒形竖置案板上，将筒口向下按成圆饼成焦饼生坯。
4. 平底锅加少量菜子油烧热，把饼坯两面煎至色金黄,再置入油锅炸熟即成。

Ingredients
500g flour, 400g beef leg meat, 1500g rapeseed oil

Seasonings
10g salt, 300g scallion, 2g MSG, 7g ginger, 10g Sichuan pepper, 15g fermented glutinous rice wine

Preparation
1. Mix the ginger and Sichuan pepper, and then mince. Finely chop the scallion. Wash the beef leg meat, remove the tendon, and mince. Add salt to the beef and mix well. Blend in 200g rapeseed oil, the fermented glutinous rice wine, chopped ginger and Sichuan pepper, MSG and scallion, then stir well to make stuffing.
2. Mix the flour with cold water to make dough, and then cover the dough with a wet towel and let stand for 30 minutes. Knead the dough into a roll. Divide the roll into small portions. Flatten each portion and make into thin strips about 1mm thick and 40cm long with a rolling pin.
3. Smear a strip of the dough with some rapeseed oil. Put some stuffing at one end of the strip, then roll up the strip from this end into the shape of a cylinder. While rolling, stretch both sides slightly. Erect the cylindrical dough on the chopping board, then press to make a round pancake.
4. Heat a little oil in a pan, and fry both sides of the pancakes till golden brown. Transfer the pancakes to a wok containing rapeseed oil, and deep-fry till cooked through.

造型美观，洁白松软，咸鲜细嫩
pleasant shape; appealing color; puffy buns; tender, aromatic and delicate stuffing

小笼包子
Steamed Buns in Small Bamboo Steamers

[原　料]

面粉500克 老面50克 半肥瘦猪肉粒500克 马蹄粒50克 冷鸡汤200克

[调　料]

食盐5克 味精2克 胡椒粉2克 料酒5克 酱油15克 芝麻油20克 白糖25克 小苏打5克 化猪油15克 姜葱水适量

[制　作]

1．面粉中加入老面、白糖、热水揉匀揉透，盖上湿毛巾后静置发酵，加入适量小苏打、化猪油揉匀，盖上湿毛巾继续静置10分钟。

2．猪肉粒、马蹄粒入盆中，加调料搅匀，再分多次加入冷鸡汤，搅打至肉粘稠成馅心。

3．面团搓成圆条，再扯成剂子，取面剂按成圆皮，放上馅心，包成细褶纹包子，入笼用旺火蒸约12分钟至熟即成。

Ingredients

500g flour, 50g sourdough, 300g hot water, 500g pork mince (with both fatty and lean meat), 50g chufa tuber (finely chopped), 200g cold chicken stock

Seasonings

5g salt, 2g MSG, 2g ground white pepper, 5g Shaoxing cooking wine, 15g soy sauce, 20g sesame oil, 25g sugar, 5g baking soda, 15g lard, ginger-and-garlic-flavored water

Preparation

1. Mix the flour, sourdough, sugar and hot water, and knead to make a dough. Cover the dough with a wet towel and let stand in room temperature to leaven. Add the baking soda and lard, mix evenly, cover again with the wet towel and let stand for another 10 minutes.

2. Mix the pork mince and chopped chufa tuber in a basin. Add the Seasonings and ginger-and-scallion-flavored water, stir well and add the cold chicken stock in successive portions, Whisk the mixture to make the stuffing.

3. Knead and roll the dough into a log, cut into sections and then press to shape each section into a round wrapper. Put some stuffing on the center of the wrapper, and gather up the edges to enclose the stuffing, twisting and pressing to seal. Transfer the buns to a specially-made bamboo steamer and steam over a high flame for 12 minutes.

色淡绿，皮软糯，馅心咸甜，清香可口
appealing greenish color; soft and fragrant dumplings

叶儿粑
Leave-Wrapped Rice Dumplings

[原 料]

糯米粉400克 籼米粉100克 淀粉20克 猪肉末150克 碎葱10克 碎米芽菜40克 猪油150克 食用油20克

[调 料]

白糖100克 蜜桂花15克 食盐1克 酱油3克 芝麻油1克 料酒4克 胡椒粉0.5克 味精0.5克

[制 作]

1. 糯米粉、籼米粉混合均匀，取十分之一加入适量沸水烫熟，再与剩下的混合米粉、少许猪油和清水调揉成光滑的面团。
2. 白糖与适量淀粉和匀，再加蜜桂花、猪油擦搓均匀成甜馅。猪油入锅烧至120℃时，下猪肉末炒散籽，加入食盐、酱油、料酒、胡椒粉炒香上色，续下芽菜炒香，起锅盛入盆内，加碎葱、芝麻油、味精拌匀成咸馅。
3. 芭蕉叶洗净后切成长约10厘米、宽约8厘米的长方片，入沸水中略煮后捞出刷上食用油；用米团分别包上甜馅和咸馅，搓成圆柱形，再用芭蕉叶裹好成叶儿粑生坯。
4. 将叶儿粑生坯放入蒸笼内，用旺火蒸约6~8分钟至熟即成。

Ingredients

400g glutinous rice flour, 100g rice flour, 20g cornstarch, 150g pork mince, 10g scallion (finely chopped), 40g minced yacai (preserved mustard stems), 150g lard, some banana leaves

Seasonings

100g sugar, 15g candied osmanthus flowers, 1g salt, 3g soy sauce, 1g sesame oil, 4g Shaoxing cooking wine, 0.5g ground white pepper, 0.5g MSG

Preparation

1. Mix the glutinous rice flour and rice flour. Get one tenth of the mixed flour and pour in some boiling water to cook it through. Mix the cooked flour, the rest of the flour and some lard, and knead into a smooth dough.
2. Mix the sugar with cornstarch, add the candied osmanthus flowers and lard, and blend well to make sweet stuffing. Heat some lard in a wok to 120℃, stir-fry the pork mince, add salt, soy sauce, Shaoxing cooking wine, ground white pepper, and continue to stir-fry till aromatic and browned. Add the minced yacai, and stir-fry to bring out the fragrance. Remove from fire, transfer to a big bowl, and add chopped scallion, sesame oil and MSG to make salty stuffing.
3. Rinse the plantain leaves, cut into rectangles about 10cm long and 8cm wide, blanch in boiling water, and smear with salad oil. Wrap the sweet or salty stuffing in pieces of the dough, shape into columns and wrap them up in plantain leaves.
4. Steam the leaf-wrapped dumplings in a steamer with energically boiling water over a high flame for 6 to 8 minutes.

形态美观，晶莹剔透，软糯香甜
crystal coating; glutinous, sweet and fragrant tangyuan

珍珠圆子
Pearly Tangyuan

[原　料]
　　糯米400克　鸡蛋液50克

[调　料]
　　白糖150克　猪油40克　蜜玫瑰10克　蜜樱桃10粒　淀粉100克

[制　作]
1. 取三分之二糯米用沸水煮至9成熟时起锅，沥干水分后置于盆内，趁热加入鸡蛋液、淀粉拌均成面团；余下的糯米提前浸泡约10小时，沥干水分留作裹米。
2. 白糖与少许淀粉拌匀；蜜玫瑰剁细，加少许猪油澥散，再将白糖、淀粉、蜜玫瑰和适量猪油揉搓均匀成馅心。
3. 取米面团包上馅心，封口后搓成圆形，均匀地粘上一层裹米，放入垫有湿纱布或刷油的笼内，并在每个生坯的顶部嵌上半颗蜜樱桃，用旺火蒸约10分钟至表面裹米过心发亮即成。

Ingredients
400g glutinous rice, 50g beaten egg

Seasonings
150g sugar, 40g lard, 10g candied roses, 10g candied cherries, 100g cornstarch

Preparation
1. Boil two thirds of the glutinous rice in water till just cooked. Drain and transfer to a basin. Add the beaten egg and some cornstarch to the basin while the rice is still hot. Blend well to make a dough. Soak the rest of the rice in water for 10 hours, drain and save to be used as coating rice.
2. Mix the sugar with some cornstarch and blend well. Chop the candied roses finely, add some lard and stir well. Mix the sugar, candied roses, the rest of the cornstarch and lard, and then knead to make stuffing.
3. Wrap up the stuffing with pieces of the dough, and kead into balls. Coat the balls evenly with rice and transfer to a steaming rack padded with wet cheesecloth or smeared with oil. Top each ball with half of a candied cherry. Steam over a high flame for 10 minutes till the coating rice becomes crystal clear.

色泽金黄，形态美观，松泡酥香
golden brown color; appealing appearance; puffy and savoury taste

蛋烘糕

Dan Hong Gao (Sichuan-Style Stuffed Pancakes)

[原 料]

面粉500克 鸡蛋3个 白糖100克 红糖50克 发酵蛋面浆150克 小苏打5克

[甜馅料]

花仁50克 白芝麻50克 白糖100克

[咸馅料]

半肥瘦猪肉150克 榨菜粒50克 食盐3克 味精1克 胡椒粉1克 料酒5克 酱油10克 食用油25克

[制 作]

1. 红糖、白糖溶于清水中，滤去渣子；面粉入盆，加入糖水、蛋液搅匀成较浓稠的蛋面浆，再加入发酵蛋面浆搅匀，静置发酵待用。
2. 花仁和芝麻分别炒香捣碎，与白糖拌匀成甜馅料；猪肉剁碎，入锅炒散，加入料酒、食盐、酱油炒干水分，加榨菜粒、胡椒粉、味精炒匀成咸馅料。
3. 蛋面浆略为发酵后加适量小苏打搅匀。
4. 特制小铜锅置小火上，烧至微热，用布蘸少许食用油将锅心抹一下，舀入少许蛋面浆，盖上小锅盖，烘约1分钟至蛋面浆即将凝结时揭开锅盖，将馅料（或甜或咸）放入蛋烘糕中央，立即用竹片将糕对叠成半月形，并把交口处压紧即可。

Ingredients

500g plain flour, 3 eggs, 100g white sugar, 50g brown sugar, 150g fermented batter, 5g baking soda

Sweet Stuffing

50g peanuts, 50g white sesame seeds, 100g sugar

Salty Stuffing

150g pork mince (a mixture of fat and lean meat), 50g zhacai (preserved mustard tuber, finely chopped), 3g salt, 1g MSG, 1g ground white pepper, 5g Shaoxing cooking wine, 10g soy sauce, 25g cooking oil

Preparaton

1. Dissolve the white sugar and brown sugar in water and filter to remove any dregs. Mix the flour, sugar juice and eggs in a basin to make batter. Add the fermented batter. Blend well and leave to ferment.
2. Stir-fry the peanuts and sesame seeds, and then pound with the sugar in a mortar to make the sweet filling. Stir-fry the pork mince in a wok to seperate. Add the Shaoxing cooking wine, salt and soy sauce, and then contintue to stir-fry to allow the water to evaporate. Add the zhacai, ground white pepper and MSG, then blend well to make the salty filling.
3. Add some baking soda to the batter after it has fermented. Blend well.
4. Put a little copper pan over a low flame, heat till slightly hot, and smear the inner surface with a dash of oil. Ladle some fermented batter into the pan and cover with a lid. Wait for about 1 minute till the batter sets. Remove the lid and place some filling (salty or sweet) at the center of the pancake. Fold the pancake into the shape of a semicircle and press the edges.

蛋烘糕 *Dan Hong Gao (Sichuan-Style Stuffed Pancakes)*

黄粑 ~ Brown Rice Cake Wrapped in Leaves

色泽金黄，香气浓郁，味甜滋润
brown color; fragrant smell; sweet, delectable and glutinous taste

黄粑

Brown Rice Cake Wrapped in Leaves

[原　料]
　糯米300克　大米200克

[调　料]
　红糖80克　白糖50克

[制　作]
1. 取部分洗净的大米、糯米打制成混合的米浆待用；再将剩下的糯米洗净，放入蒸笼内蒸至7~8成熟待用。
2. 将打制好的米浆与蒸好的糯米饭倒入大木盆中加入少量红糖、白糖拌匀。
3. 待米浆中的水分被糯米饭完全吸收后，将糯米饭搓打成一个个大饭团，并摔打成型。
4. 用洗净、晾干的芭蕉叶将糯米饭团依次捆扎好，入锅蒸熟即成。

Ingredients
300g glutinous rice, 200g rice

Seasonings
80g brown sugar, 50g white sugar

Preparation
1. Rinse the glutinous rice and rice in water. Mix the rice and some glutinous rice, grind to make rice milk. Steam the remaining glutinous rice till just cooked.
2. Mix the rice milk with the steamed glutinous rice in a wooden basin, blend well and add the brown sugar and white sugar. Stir well.
3. Wait till the milk is fully absorbed by the glutinous rice, and then knead the glutinous rice into balls. Shape the balls into rectangular cakes by dumping them with force onto a board repeatedly.
4. Rinse the plantain leaves and wrap the rice cakes up. Enlace each of the cakes tightly and steam them till cooked through.

面条爽滑筋道，咸鲜香辣，甜酸适口
smooth and springy noodles; a mixture of salty, pugent, sweet and sour taste; aromatic smell

鸡丝凉面

Cold Noodles with Shredded Chicken

[原　料]
面条500克　绿豆芽150克　熟鸡肉丝200克　熟菜油30克

[调　料]
复制红酱油100克　酱油100克　红油辣椒100克　芝麻酱50克　花椒油5克　蒜泥50克　芝麻油5克　食醋60克

[制　作]
1. 锅置旺火上，入清水烧沸，将面条下锅煮至断生后捞出，沥干水分后摊开，洒少量熟菜油后抖散、晾冷；将调料混合均匀成味汁。
2. 绿豆芽入沸水中煮熟，捞出晾冷后放入碗内，加凉面和味汁，撒上鸡丝即成。

Ingredients
500g noodles, 150 mung bean sprouts, 200g precooked chicken, shredded, 30g precooked rapeseed oil

Seasonings
100g concocted soy sauce, 100g soy sauce, 100g oil-infused chili flakes, 50g sesame paste, 5g Sichuan pepper oil, 50g garlic (finely chopped), 5g sesame oil, 60g vinegar

Preparation
1. Heat water in a wok over a high flame, bring to a boil, add noodles and continue to boil till just cooked. Remove, drain and spread out. Sprinkle the noodles with precooked rapeseed oil, mix well to prevent the noodles from sticking together, and cool. Mix the Seasonings to make seasoning sauce.
2. Boil the mung bean sprouts till cooked through, remove, cool and transfer to a bowl. Add the noodles and the seasoing sauce. Top the noodles with the chicken shreds.

军屯锅盔
Juntun Pancakes

色泽金黄，质地酥脆，香鲜爽口

golden-brown color; crunchy, crispy crust; savory and aromatic taste

[原 料]

面粉300克 猪肉250克 猪油100克 食用油100克

[调 料]

香葱30克 姜米30克 花椒粉5克 味精3克 食盐5克 胡椒粉1克 酱油8克

[制 作]

1. 面粉加入少许食盐、猪油和适量清水调成偏软的面团，盖上湿毛巾饧面30分钟。
2. 猪肉剁细，加入调料拌和均匀成馅心。
3. 将饧好的面团搓成长条，再扯成大小均匀的面剂，将其擀开并涂上肉馅，先卷成圆筒，再压扁擀成小圆饼即成生坯。
4. 平底锅在火上炙好后加入少许食用油烧热，放入生坯煎至两面均呈金黄色，再将其放入炭炉中烘烤至酥脆即成。

Ingredients

300g flour, 250g pork, 100g lard, 100g looking oil

Seasonings

30g scallion, 30g ginger (finely chopped), 5g ground roasted Sichuan pepper, 3g MSG, 5g salt, 1g ground white pepper, 8g soy sauce

Preparation

1. Mix the flour with some salt, lard and water to make a soft dough. Cover with a wet towel and let stand for 30 minutes.
2. Chop finely the pork, and mix with the Seasonings to make stuffing.
3. Knead the dough into a roll, tear into small portions, roll with a rolling pin into long slices and smear with the stuffing. Roll up the slices and flatten into round pancakes with the rolling pin.
4. Heat a pan over a flame, add some salad oil and wait till the oil is hot. Fry the pancakes till golden brown on both sides and then bake in a coal stove till crispy.

糯米软糯，香甜可口
glutinous and soft rice; sweet and luscious taste

三大炮
Three Cannonshots (Sweet Rice Buns)

[原 料]
糯米1000克

[调 料]
红糖150克 芝麻50克 黄豆250克

[制 作]
1. 糯米洗净，用清水浸泡12小时，淘洗后入蒸笼中用大火蒸熟（中途洒1～2次水），翻倒在木桶内，掺沸水适量，用盖盖上，待水分进入米内后，用木棒舂蓉成坯料。
2. 将红糖、清水放入锅中用小火熬成糖汁。
3. 分别把芝麻、黄豆炒熟，再磨成细粉。
4. 将坯料分成10份，再把每份分成3块捏成团，连续甩向木盘，发出三声响后弹入装有黄豆粉的簸箕内，均匀裹上黄豆粉后装碗，淋上红糖汁，撒上芝麻面即成。

Ingredients
1000g glutinous rice

Seasonings
150g brown sugar, 50g sesame seeds, 250g soy beans

Preparation
1. Wash the glutinous rice and soak in water for 12 hours. Rinse, transfer to a steaming rack and steam over a medium-high flame till fully cooked (spray the rice with water once or twice during the steaming process). Transfer the rice to a wooden pail, add some boiling water, cover with the lid and then wait till the water is absorbed by the rice. Pummel and grind the rice with a wooden pestle.
2. Add the brown sugar and some water into a wok and simmer over a low flame to make sugar juice.
3. Roast the sesame seeds and soy beans respectively, and grind into fine powder.
4. Divide the ground rice into 10 portions. Then divide each of the 10 portions into three parts. Knead the three parts into three balls, and fling the balls into a wooden tray. The three balls, after hitting the tray with three bangs, would land in a bamboo basket containing the soy bean powder. Coat the balls with soy bean powder, transfer to a bowl, and drizzle with sugar juice and sesame seed powder.

赖汤圆 Lai's Tangyuan (Sweet Rice Dumplings)

皮薄馅多，软糯细腻，滋润香甜
thin wrapper and rich fillings; soft, glutinous and sweet taste

赖汤圆
Lai's Tangyuan (Sweet Rice Dumplings)

[原 料]

糯米粉团200克 黑芝麻70克 白糖粉300克 熟面粉50克 化猪油200克

[调 料]

白糖、芝麻酱适量

[制 作]

1. 黑芝麻用中小火炒香、炒熟，压成粉，加入白糖粉、熟面粉、化猪油揉匀，再切成1.5厘米见方的小块成馅心。
2. 糯米粉团分成数份，分别装入馅心，包捏成圆球状成汤圆生坯。
3. 锅置旺火上，掺水烧沸，入汤圆生坯煮熟后捞出，放入碗中。食用时配上白糖、芝麻酱小碟蘸食。

Ingredients
200g glutinous rice flour dough, 70g black sesame seeds, 300g sugar, 50g pre-cooked flour, 200g lard

Seasonings
sugar and sesame paste to taste

Preparation
1. Roast the black sesame seeds over a medium-low flame till aromatic and fully cooked, crush into powder and mix with sugar, pre-cooked flour and lard. Blend well, knead, and cut into 1.5cm^3 cubes to serve as stuffing.
2. Divide the dough into several portions to serve as wrappers. Enclose some stuffing in a wrapper, knead and roll into a ball to make Tangyuan.
3. Heat water in a wok over a high flame, bring to a boil and roll in the Tangyuan. Boil till cooked through, remove, and transfer to a bowl. Serve with sugar and sesame paste on saucers for dipping.

典故
成都名小吃。创始人赖源鑫以卖汤圆为业，日久出名，人们以"赖汤圆"称之。后其于成都春熙路开店经营汤圆，店名亦为"赖汤圆"。

Allusion
It's a famous snack in Chengdu. The Tangyuan sold by Lai Xinyuan was delicious and gradually became famous. People began to call it Lai's Tangyuan to distinguish it from others. Later Lai Xinyuan started a business on Chunxi Road, and he named his restaurant Lai's Tangyuan.

面筋道滑爽，汤鲜美适口
smooth and springy pasta; delicate and savoury soup

铺盖面
Sheet Pasta with Topping

[原　料]
高筋面粉200克　普通面粉300克　干豌豆100克　鲜汤200克

[调　料]
食盐6克　味精5克　胡椒粉2克　酱油10克　化猪油少许　葱花少许

[面　臊]
红烧排骨、炒鸡杂、红烧肥肠、红烧牛肉等

[制　作]
1．高筋面粉、普通面粉加入食盐和适量清水调制成质地偏硬的面团，再盖上湿毛巾饧面30分钟。
2．干豌豆入锅煮至软烂时取出待用。
3．食盐、味精、胡椒粉、酱油、化猪油、豌豆、葱花和适量鲜汤放入碗中定味。
4．锅置火上，加入清水烧沸，取一小块面团用手拉扯成圆形的薄面皮，放入锅中煮熟后捞出，放入碗中，浇上面臊即成。

Ingredients
200g bread flour, 300g all-purpose flour, 100g dried peas, 200g everyday stock,

Seasonings
6g salt, 5g MSG, 2g ground white pepper, 10g soy sauce, lard, chopped scallion

Topping
braised spareribs, beef, chicken offal, pork intestines

Preparation
1. Mix the bread flour, all-purpose flour, some salt and water. Blend well to make a dough. Cover the dough with a wet towel and leave to rest for 30 minutes.
2. Boil the peas in water till soft and cooked through. Remove.
3. Mix salt, MSG, ground white pepper, soy sauce, lard, peas, scallion and stock in a bowl.
4. Put a wok over a flame, add water and bring to a boil. Get a small portion of the dough and make it into a thin round slice with hands. Boil the slices in water till cooked through, remove into the bowl and pour different toppings over the pasta.

Noodles with Red-Braised Beef Topping
红烧牛肉面

入口爽滑，鲜美微辣，醇香浓厚

salty and slightly hot taste; aromatic and savory beef

[原　料]

面条500克　牛肉块250克　鲜汤适量

[调　料]

生姜50克　葱12克　料酒5克　糖色5克　花椒5克
酱油100克　豆瓣100克　食盐5克　芝麻油5克
味精3克　葱花12克　香料包1个

[制　作]

1. 锅置中火上，加鲜汤烧沸，入牛肉煮沸，加食盐、料酒、葱、生姜、豆瓣、花椒、香料包、糖色烧至牛肉软烂，去掉生姜、葱、香料包、花椒成面臊。
2. 酱油、葱花、芝麻油、味精放入碗内兑成味汁。
3. 锅置火上，入清水烧沸，下面条煮熟后捞出，放入味汁碗内浇上面臊即成。

Ingredients

500g noodles, 250g beef chunks, everyday stock

Seasonings

50g ginger, 12g scallion, 5g Shaoxing cooking wine, 5g caramel color, 5g Sichuan pepper, 100g soy sauce, 100g chili bean paste, 5g salt, 5g sesame oil, 3g MSG, 12g finely chopped scallion, 1 packet of mixed herbal spices

Preparation

1. Heat the stock in a wok over a medium flame, bring to a boil, add the beef and bring to a second boil. Add the salt, Shaoxing cooking wine, scallion, ginger, chili bean paste, Sichuan pepper, the packet of spices and caramel color. Continue to boil till the beef is soft and cooked through. Remove and discard the ginger, scallion, packet of spices and Sichuan pepper.
2. Combine the soy sauce, chopped scallion, sesame oil and MSG in a serving bowl.
3. Heat water in a wok, bring to a boil, boil the noodles till cooked through, remove, and transfer to the serving bowl. Top the noodles with the braised beef made in Step 1.

粉条滑爽，咸鲜酸辣，鲜香可口
smooth sweet potato noodles; appetizing sour-and-spicy taste

酸辣粉

Hot-and-Sour Sweet Potato Noodles

[原　料]
　　红薯粉条300克　豌豆苗100克

[调　料]
　　食盐15克　味精2克　胡椒粉2克　花椒粉5克
　　酱油75克　醋50克　辣椒油75克　化猪油100
　　克　碎米芽菜50克　芹菜末50克　葱花50克

[制　作]
1. 红薯粉条加热水泡发后换清水继续浸泡；调料装入小碗中。
2. 汤料锅置火上烧沸，取红薯粉条少许装入竹漏中，放入汤中煮约3分钟，加少许豌豆苗略煮后装入盛有调料的碗中拌匀即成。

Ingredients
300g dried sweet potato noodles, 100g pea vine sprouts

Seasonings
15g salt, 2g MSG, 2g ground white pepper, 5g ground roasted Sichuan pepper, 75g soy sauce, 50g vinegar, 75g chili oil, 100g lard, 50g yacai (preserved mustard stems, minced), 50g celery (finely chopped), 50g scallion (finely chopped)

Preparation
1. Soak the sweet potato noodles in hot water till soft, and then transfer to cold water. Mix the Seasonings in a small bowl.
2. Heat stock in a wok. Put some noodles in a slotted bamboo ladle, transfer the ladle into the stock and make sure the noodles are submerged. Boil the noodles for about 3 minutes, add pea vine sprouts to the ladle, boil briefly and transfer to the bowl.

形似鱼网，饺底金黄酥脆，饺面柔软油润，饺馅鲜嫩多汁
fishing-net-shaped coating; golden brown color; soft and smooth wrappers; juicy and aromatic stuffing

一品锅贴
Deluxe Fried Dumplings

[原　料]
面粉500克 猪肉500克 淀粉浆400克 食用油100克

[调　料]
鸡汤200克 味精5克 食盐5克 胡椒粉2克 料酒10克 白糖10克 芝麻油10克 葱姜水50克

[制　作]
1. 面粉加入热水调制成三生面团，并反复揉搓均匀，盖上湿毛巾饧面。
2. 猪肉铰细后加入食盐、料酒、味精、胡椒粉、白糖拌匀，再加入葱姜水、鸡汤、芝麻油搅拌为呈粘稠糊状的馅心。
3. 面团搓成圆柱形长条，再切成剂子，擀成直径约8厘米的圆皮，装上馅心并捏成月牙形生坯。
4. 平底锅炙好后加入少许食用油烧热，放入饺坯煎至底部冒油泡时，撒入淀粉浆并盖上锅盖，续煎至饺底色泽金黄即成。

Ingredients
500g flour, 500g pork, 400g cornstarch batter (100g cornstarch dissolved in 300g water), 100g looking oil

Seasonings
100g chicken stock, 5g MSG, 5g salt, 2g ground white pepper, 10g Shaoxing cooking wine, 10g sugar, 10g sesame oil, 50g ginger-and-scallion-flavored water

Preparation
1. Mix the flour with hot water to make a dough, and knead till smooth. Cover the dough with a wet towel and let stand for a while.
2. Mince the pork, add the salt, Shaoxing cooking wine, MSG, ground white pepper and sugar, and mix well. Blend in the ginger-and-scallion-flavored water and chicken stock, mix well to make a paste, and add the sesame oil to make the stuffing.
3. Roll the dough into a cylindrical strip, cut into small portions, and flatten them into round wrappers about 8cm in diameter with a rolling pin. Fill the wrapper with stuffing and press the edges to make dumplings shaped like a first quarter moon.
4. Heat a pan over a flame, add some oil and go on heating till hot. Add the raw dumplings and fry till the the oil bubbles. Sprinkle the cornstarch batter over the dumplings, cover and continue to fry till the bottoms of the dumplings become golden brown.

Sichuan (China) Cuisine in Both Chinese and English

川菜
(中英文标准对照版)

1. 川菜特色调味品
Featured Seasonings

2. 川菜烹饪术语
Terms

附录
Appendix

1. 川菜特色调味品 Featured Seasonings

郫县豆瓣

郫县豆瓣又称辣豆瓣、辣豆瓣酱，是选用新鲜红辣椒以传统工艺精酿发酵而成。在四川所有的豆瓣酱中，首推郫县酿造的"郫县豆瓣"。郫县豆瓣由郫县人陈守信始创于清朝康熙年间，至今已愈三百年历史。其特色是红褐色、略油润、有光泽，有酱酯香和辣香，味鲜辣，瓣粒酥脆化渣、回味较长、粘稠适度。用它炒菜、制作火锅分外提色增香，被誉为"川菜之魂"。

Pixian Chili Bean Paste

Chili bean paste is made by fermenting fresh red chili peppers by way of traditional technology. Among all the brands in Sichuan, Pixian Chili Bean Paste made in Pixian is the most renowned. Started by Chen Shouxin in Pixian during the Emperor Kangxi's reign of the Qing Dynasty, Pixian Chili Bean Paste has a history of over 300 years. It features red and lustrous color, spicy and aromatic taste, soft and melting beans, and lingering fragrance and moderate viscosity. The chili bean paste is often used for flavor stir-fried dishes or hot pot, and referred to as "the Soul of Sichuan Cuisine".

小米辣椒

小米辣椒属于半驯化小果型特色辣椒。单产高、果型小、辣味强、香味浓郁、口感良好，是理想的辣味调料。小米辣椒主要产于云南境内，也有少量产自四川，与子弹头辣椒、二荆条辣椒相比，辣味更为突出，烹调上多用于调味和制作成蘸碟。随着人们生活质量的提高，小米辣椒以其果型新奇美观、辣味独特，既能烹制美味佳肴，又可增进食欲、促进血液循环和防感冒、御风寒而深受消费者的欢迎，在现代川菜中被广泛使用。

Birdseye Chili Peppers

Birdseye chili peppers, a kind of half-acclimatized and small-fruit pepper, are often used as a seasoning for its strong pungency. They are mainly found in Yunnan. Some of them are grown in Sichuan. Compared with Zidantou chili peppers and Erjingtiao chili peppers, birdseye chilies have an even stronger hot taste and are often used for flavor dishes or dipping sauce. With the improvement of people's lives, birdseye chili peppers are becoming increasingly popular for their unique taste, peculiar appearance, and such medical benefits as giving appetite, preventing cold and reducing the risk of suffering from cardiovascular diseases. They are widely used in modern Sichuan cuisine.

泡辣椒

泡辣椒又称鱼辣椒，是由新鲜红辣椒泡制而成。一般选用二荆条辣椒或朝天椒、小米辣椒来泡制。四川是全国辣椒的主要产区之一，所产的二荆条辣椒质量最好。其色鲜红、质嫩，椒角细长、肉厚籽少，辣味适中，香味与辣味俱佳，维生素含量高，最适宜做豆瓣、泡椒和辣椒油。

由于泡辣椒在泡制过程中产生了乳酸，用来烹制菜肴，会使菜肴具有独特的香气和味道，还能去除鱼类的腥味，是川菜中烹制鱼肴和鱼类火锅等的主要调料。

Pickled Chilies

Pickled chilies, also called fish chilies, are made by pickling fresh chili peppers. The varieties often preserved are Erjingtiao chili peppers, birdseye chili peppers and upward-pointing chili peppers. Sichuan is one of the main pepper-producing regions in China, and the Erjingtiao chili peppers, a kind of long thin chili produced here are of excellent quality with red color, tender pulp, fewer seeds, pleasant fragrance and mildly pungent taste. Erjingtiao chili peppers are also rich in vitamins and often used to make chili bean paste, pickles and chili oil.

Due to the fact that lactic acid is produced during the pickling process, pickled chilies are usually used to render dishes a special taste and fragrance. Sichuan cuisine mainly employs the pickled chilies to remove the unpleasant smell of the fish when preparing fish dishes or hot pot.

野山椒

野山椒属指天椒，主要用于菜肴烹调、配菜的佐料。野山椒既可鲜食，也可干食，还可以深加工制成罐头食品。四川地区多以泡制为主，广泛用于川菜的调味。在烹制川味菜肴时加入适量泡野山椒，不仅可去腥除臊，还可将原本平淡无味的原料变成鲜美可口的佳肴。

Tabasco Chili Peppers

Tabasco chili peppers, a upward-pointing chili, are mainly used in cooking as a seasoning. They can be eaten fresh, canned or dried. The canned green tabasco chili peppers have a strong footing in the market. The Sichuanese often pickle them before applying them to dishes as the pickled tabasco chili peppers not only help to remove the unpleasant smell of some ingredients but also have the magic of turning plain dishes into delicacies.

汉源花椒

汉源花椒产自四川省汉源县，历史悠久，唐代列为贡品，又称"贡椒"。其色丹红，颗粒饱满，粒大肉厚，油重质佳，麻香味浓，深受消费者欢迎。

花椒在烹调中主要呈现麻味，同时还有激发食欲、压异、增香的作用，在川菜烹调中使用广泛，不仅直接用于火锅、水煮、干煸和烹煮肉类等菜肴，还可制成椒盐味碟、椒麻糊等复合调味品使用。

Hanyuan Sichuan Pepper

Hanyuan Sichuan Pepper refers to the Sichuan Pepper that is grown in Hanyuan County. It has a fairly long history and is also called "Tribute Pepper" for it used to be presented to the royalty during the Tang Dynasty. This species features purple red color, plump fruit, big size, rich oil and strong tingling taste, and is popular among consumers.

Sichuan pepper, with its tingling flavor, is widely used in Sichuan cuisine to stimulate the appetite, remove the unpleasant smell and add to the flavors. It can not only be used in preparing hot pot, meat dishes, dry-fried dishes, boiled dishes in chili sauce and hot-and-spicy dishes, but also adopted when making salty-and-tingling dipping sauce and such compound seasonings as Jiaoma paste.

青花椒

青花椒又名油椒、香椒子、藤椒，是一种有独特香气和味道的植物，其果油多有光泽，其味芬芳易挥发，口感香麻，是重要的油料、香料。有鲜品和干制品两种，还可加工制作成藤椒油，具有调味、健脾、祛风散寒之功效。藤椒用于烹调，口味清爽，麻香浓郁，麻味绵长，比花椒更香、更麻，广泛用于现代川菜调味。

Green Sichuan Pepper

Green Sichuan pepper has a unique fragrance and taste and is mainly used as an oil plant or a spice. It is available in the market either fresh or dried. Oil extracted from the pepper is also used to season food. Compared with common Sichuan pepper, green Sichuan pepper, with a stronger fragrance and a more tingling taste, is more refreshing and appetizing. It's widely used in modern Sichuan cuisine.

宜宾芽菜

宜宾芽菜是四川省宜宾市的特产，有着悠久的历史。它是用青芥菜去叶剖丝，适度晾晒后拌入食盐、红糖，再加入香料后装坛腌储而成，是四川省著名的传统"四大腌菜"之一，具有香、甜、脆、嫩、鲜等特点，在川菜烹调中多作调辅料使用。

Yibin Yacai

Yibin yacai (preserved mustard stems) is peculiar to Yibin, Sichuan. It has a long history and is made with mustard stems. The leaves of the mustard plants are first removed, and the stems are then cut into long slivers, air-dried, mixed well with salt, brown sugar and spices, and finally sealed in earthen jugs to preserve. It ranks among the Four Most Renowned Preserved Vegetables in Sichuan and has the characteristics of being aromatic, sweet, crispy, tender and savory. Yacai usually serves as a condiment or complementary ingredient in cooking.

四川榨菜

榨菜是中国的特产，为腌菜中的佳品，在国际市场上与欧洲酸菜、日本酱菜齐名，具有鲜、香、脆、嫩的独特风味，被誉为世界三大名腌菜之一。中国榨菜生产遍及全国14个省、市，其中四川的产量位居第一，质量为全国之冠，是传统的出口商品之一，行销日本、东南亚和欧美等10多个国家和地区。

Sichuan Zhacai

Zhacai (Preserved mustard tuber) is peculiar to China, and enjoys the same fame as European pickles and

Japanese tsukemono in the international market. It has the features of being savoury, aromatic, crispy and tender and is produced in 14 provinces of China, of which Sichuan boasts the largest output and highest quality. Sichuan zhacai has long been exported to Japan, south-eastern Asia and over 10 countries in Europe and America.

南充冬菜

南充冬菜是四川省南充市的特产，为四川著名的传统"四大腌菜"之一。它是用芥菜的嫩尖经独特工艺腌制发酵后制成，质地嫩脆，色彩黑褐，有光泽，香气浓郁，风味鲜美独特。

南充冬菜富含氨基酸、乳酸、蛋白质、维生素和多种微量元素，有开胃健脾、增进食欲、增强人体机能等功效，常用作炒、炖、蒸等菜肴和汤品、面食的调料。

Nanchong Dongcai

Nanchong dongcai, one of the four "Most Renowned Preserved Vegetables", is made by pickling and fermenting the tender sprouts of mustards in a unique way. It is tender, crispy, aromatic, and delicate with dark brown color.

Nanchong Dongcai is rich in amino acid, lactic acid, protein, vitamins and microelements, and is often used as a seasoning when preparing stir-fried dishes, steamed dishes, stew, soup and noodles.

大头菜

大头菜是四川著名的传统"四大腌菜"之一。它是将根用芥菜经风吹日晒后腌制加工而成，色泽淡黄，质地软中带脆，香气浓郁，咸淡适宜。在川菜烹调中，多作为鱼类火锅、蒸菜、拌菜、面食、佐餐等的调料。

kohlrabi

Kohlrabi is one of the "Four Most Renowned Preserved Vegetables" in Sichuan. Mustard is first air-dried and then salted and processed for preservation. It is of light brown color, tender but crispy texture, strong aroma and mild taste. It is often used as a seasoning when preparing fish hot pot, steamed dishes, salad, noodles and side dishes, etc.

自贡井盐

自贡井盐有1900多年的历史，产自有"中国盐都"之称的四川省自贡市。它是用深藏于地下近千米的盐卤熬制而成，氯化钠的含量高达99%以上，杂质甚微，味道纯正，品质优异，是川菜烹调用盐和制作泡菜的首选。

Zigong Well Salt

Zigong well salt, with a history of more than 1900 years, is produced in Zigong City of Sichuan province, which is called the Capital of Salt in China. It is made by boiling down the brine taken from about 1000 meters underground. The brine contains over 99% sodium chloride and few impurities. The salt, with orthodox taste and good quality, is the first choice for making pickles and is often used in cooking.

中坝酱油

中坝酱油是四川省江油市的特产。它以优质大豆为主料，经日晒夜露、天然发酵而成，色艳汁稠，咸甜适度，醇香浓郁，耐久存，适用于烹制各种冷热菜肴，尤其适合日常炒菜，属上乘调味佳品。

Zhongba Soy Sauce

Zhongba soy sauce, a local specialty of Jiangyou, Sichuan, is made with high-quality soy beans, which are naturally fermented by exposing in the sun and during the night. The thick sauce features pleasant color, mildly salty and sweet taste, and lingering smell. It can be preserved for a long time, and is often used in both hot and cold dishes.

阆中保宁醋

阆中保宁醋产于四川省阆中市，是全国四大名醋之一。它以历史悠久、质优味美、营养丰富、防病治病而著称于世，曾在1915年巴拿马"太平洋万国博览会"上荣获金奖。保宁醋呈黄棕色，汁稠挂碗，酸味柔和，醇香回甜，久存不腐且香气愈浓，是优质的烹饪及佐餐调味料。

川菜特色调味品 Featured Seasonings

Langzhong Baoning Vinegar

Langzhong Baoning Vinegar, one of the "Four Most Renowned Vinegars", is famous for its long history, excellent quality, orthodox taste, rich nutritious elements and medical benefits. At the 1915 Panama Pacific International Exposition, Baoning Vinegar won the gold medal. This brown vinegar is thick, fragrant, and of mild sour and slightly sweet taste. This choice seasoning can be preserved for a long time.

新繁泡菜

新繁泡菜产自中国泡菜之乡——四川省成都市的新繁古镇。它是精选时令蔬菜经特制的盐水泡制而成，色泽鲜美、脆嫩芳香、解腻开胃。现有甜酸味、咸酸味、红油辣味、日式风味、韩式风味等5大系列200多个品种，其代表产品有泡青菜、泡萝卜、泡辣椒等，是川菜烹调、佐餐的常用调辅料。

Xifan Pickles

Xinfan pickles refer to those made in the ancient town Xinfan, Chengdu. Seasonal vegetables are first selected, and then pickled in specially-made brine. They feature bright color, fragrant smell and crispy texture, and serve to remove the grease and stimulate the appetite. At present there are over 200 kinds under five series according to the taste: sweet-and-sour, salty-and-sour, chili-oil-flavored, Japanese-style and Korean-style. Pickled green vegetables, pickled radish, pickled chili peppers are typical of them. Xinfan Pickles are often used as a seasoning or a side dish in Sichuan cuisine.

潼川豆豉

潼川豆豉产于四川省三台县，因三台古称潼川府，故习惯称为潼川豆豉。它是选用优质黑色大豆酿制而成，颗粒松散、色彩黝黑、有光泽，清香回甜，滋润化渣。用以烹调炒、煸、拌等菜肴和制汤，最能体现川菜风味，是川菜大师们专选的调味品之一。

Tongchuan Fermented Soy Beans

Tongchuan fermented soy beans are made in Santai County of Sichuan Province, which used to be called Tongchuan in ancient times. It is made with black soy beans and features pleasant color, fragrant smell, salty and slightly sweet taste, and delicate texture. It is one of the seasonings peculiar to Sichuan cuisine, and is often added when preparing stir-fried dishes, dry-fried dishes, salad and soup.

唐场豆腐乳

唐场豆腐乳产于四川省大邑县唐场镇。它是将豆腐坯切块后，放入有适当温度的环境中发酵、腌制而成，味鲜可口，细嫩化渣，清香回甜，享有较高的声誉。用以佐餐、调味，能增进食欲、帮助消化。

Tangchang Fermented Tofu

Tangchang fermented tofu is made in Tangchang Township, Dayi County, Sichuan Province. Tofu is first cut into chunks, and then fermented and salted in appropriate temperature. It is delicate, fragrant, tender and melting in the mouth, and can be used as a side dish or a seasoning. It has the benefits of stimulating the appetite and helping digestion.

醪糟

醪糟又称米酒、酒粮、甜酒等，是由糯米或大米经过发酵而制成，含有丰富的营养成分、大量水分和少量酒精，烹调中主要用其汁水部分，称之为醪糟汁。醪糟汁在川菜烹调中多用于火锅、红烧、干烧、糟醉、小吃等菜点的调味，有压腥、去异、增香和助消化、增进食欲等作用。

Glutinous Rice Wine

Glutinous rice wine is made by fermenting rice or glutinous rice and contains rich nutritious elements, large quantities of water and small quantities of alcohol. It is its juice that is often employed when cooking hot pot, red-braised dishes, dry-braised dishes, wine-flavored dishes as well as some snacks so as to remove the unpleasant smell, add to the flavor, help digestion, and stimulate the appetite.

2. 川菜烹饪术语 Terms

清汤 consomme

是用鸡、鸭、猪瘦肉、排骨加水，经小火长时间熬制后取汤，加食盐、胡椒粉调味，再加入肉蓉使其凝固，取澄清液过滤而成。汤汁清澈见底，味道咸鲜醇厚。

Simmer chicken, duck, lean pork and pork ribs in water to make stock. Season with salt and pepper, and then add minced meat to absorb the floating particles or sediments. Filter to get clear stock which is pellucid, delicate and savory.

奶汤 milky stock

是用鸡、鸭、猪肘、猪骨、猪肉加水，用中火熬制而成。汤汁呈乳白色，形如乳汁，故名。

Simmer chicken, duck, pork knuckle, pork bones and pork over a medium-low flame till the soup becomes milky.

鲜汤 everyday stock

是用猪肉、鸡等加水熬制而成。

Simmer chicken or pork in water over a low flame.

鲜 xian

在中国，鲜具有广泛的内涵，既指新鲜，也指鲜味。鲜味是一种复杂的味感，主要呈味成分有核苷酸、氨基酸（谷氨酸钠）、有机酸等，主要调味料有味精和鸡精。

其中，味精主要是用大米、玉米等粮食或糖蜜，采用微生物发酵的方法提取而成。它的主要成分是谷氨酸钠，是氯化钠的助味剂，无氯化钠则感觉不出鲜味，最宜与咸味配合成咸鲜味。

鸡精是以谷氨酸钠、食用盐、呈味核苷酸二钠（一种增鲜剂）、鸡肉与鸡骨的粉末或其浓缩提取物等为基本原料，添加或不添加香辛料和食用香精等增味、增香剂经混合、制粒、干燥加工而成，是具有鸡肉的鲜味、香味的复合调味料。

In Chinese, the word "xian" is abundant in meaning, which might refer to the freshness of ingredients or a flavor. Xian is a compound flavor, which is primarily rendered by such substances as nucleotide, aminophenol (sodium glutamate) and organic acid. MSG and chicken essence granules are the two main seasonings to enhance the xian of dishes.

MSG is extracted from rice, corn or molasses by using microbial fermentation. Its main ingredient is sodium glutamate, which would lose its flavor-enhancing function without sodium chloride. MSG works best with salt, and they combine to result in the taste of salty xian.

Chicken essence granules uses sodium glutamate, salt, disodium nucleotide and ground chicken and ground chicken bones as the main ingredients. The ingredients might be spiced with taste-enhancing or smell-enhancing additives before being granulated and dried. The seasoning is characteristic of chicken-flavor xian.

醇厚 Chun Hou

在烹饪上指菜肴或汤汁爽适甘厚、纯正柔和，有醇香和酯香，味厚绵长。

The two characters "Chun Hou" in Chinese refer to the mild, mellow and lingering taste of soups or dishes, mainly rendered by ester.

鸡油 chicken oil

是用鸡腹内脂肪蒸制而成。

The oil is made by steaming chicken belly fat.

油发 oil-soaking

系一种涨发方法。是将干货原料置于油中，经加热后使物体内部水分蒸发，形成空洞结构而蓬松涨大，再用水泡软而使原料可用。

Soak dried ingredients in oil, heat to evaporate the inner water content so that the ingredients become swollen because inner cavities are formed. Then soak the ingredients in water till soft.

拌 Sichuan-style salad

是将生原料或熟原料加工成型，再用调味品拌制成菜的烹调方法。

Cut ingredients (raw or cooked) into wanted shapes, add condiments and stir well.

糖色 caramel color

冰糖或白糖入锅，用中小火炒制成棕红色，加入少许水制作而成。

Heat sugar or rock sugar in a wok over a medium-low flame till it has melted and become dark brown. Add some water to make the caramel color.

椒麻糊 Jiaoma paste

又称葱椒糊，是用葱青叶和花椒按照8～10∶1的比例剁细成末，再加入少许100℃食用油制作而成。

Mix eight to ten portions of scallion leaves with one portion of Sichuan pepper, chop finely and add some 100℃ oil.

煳辣味型 Hula flavor

是四川菜常用的复合味之一。主要是用干辣椒、花椒入油锅中炸至表面略焦、色呈棕褐时加白糖、醋等调味料调制而成，味道麻辣不燥、荔枝味突出，咸鲜香醇厚。

It is one of the widely-used compound flavors of Sichuan cuisine. Fry dried chilies and Sichuan pepper in oil till their surfaces become a little burnt and dark brown. Add sugar, vinegar and other condiments to season. The flavor is spicy and moreish with slightly sweet and sour taste.

卤水 Sichuan-style broth

是四川菜点中常用的一种调味卤汁，系用花椒、八角、陈皮、桂皮、山柰、小茴香、甘草、草果、生姜、葱、生抽、老抽、冰糖及鲜汤等多种原材料，用小火煮制数小时、待香味浓郁后而成。

It is one of the widely-used seasoning broth in Sichuan cuisine. Its ingredients include Sichuan pepper, star anise, tangerine peel, cinnamon, sand ginger, fennel seeds, liquorice, tsaoko amomum, ginger, scallion, soy sauce, rock sugar and stock, etc, which are simmered over a low flame for hours till aromatic.

Sichuan (China) Cuisine in Both Chinese and English

川菜

(中英文标准对照版)